Saving Freedom

Truman, the Cold War, and the Fight for Western Civilization

Joe Scarborough

HARPER LARGE PRINT

An Imprint of HarperCollinsPublishers

HarperCollins books may be purchased for educational, business, or sales promotional use. For information, please e-mail the Special Markets Department at SPsales@harpercollins.com.

FIRST HARPER LARGE PRINT EDITION

ISBN: 978-0-06-302971-2

Library of Congress Cataloging-in-Publication Data is available upon request.

20 21 22 23 24 LSC 10 9 8 7 6 5 4 3 2 1

Saving Freedom

ALSO BY JOE SCARBOROUGH

*The Right Path: From Ike to Reagan, How Republicans
Once Mastered Politics—and Can Again*

*The Last Best Hope: Restoring Conservatism
and America's Promise*

*Rome Wasn't Burnt in a Day: The Real Deal on
How Politicians, Bureaucrats, and Other Washington
Barbarians Are Bankrupting America*

To the memory of Dr. Brzezinski
and all those public servants
who dedicated their lives to liberating Europe
from the scourge of Soviet communism

I have sworn upon the altar of God, eternal hostility against every form of tyranny over the mind of man.

—THOMAS JEFFERSON

In the long run the strength of our free society, and our ideals, will prevail over a system that has respect for neither God nor man.

—HARRY TRUMAN

Contents

Introduction
A Strange Little Man

The portraits of Harry Truman and Ronald Reagan hung in my congressional office throughout the time I served in Congress. The image of Reagan surprised few, since most viewed me as a fire-breathing Republican whose small-government crusades on the House floor and in GOP caucus meetings seemed extreme even to Republican peers. Besides, Ronald Reagan was a unifying figure who left office with high approval ratings and an abundance of goodwill. But Truman's portrait, the same that graces the cover of this book, surprised visitors. In his time, Harry Truman was not a beloved figure even among Democrats. Unlike the Gipper, Truman departed the White House with historically low levels of support. Still, I considered myself a conservative more than a Republican, and in the Scarborough household, being conservative was defined primarily by one's status as a Cold War hawk. Truman fit that bill and was, in fact, the first Cold Warrior in the White House.

When I came to Washington in 1994, the United States was still basking in the glow of its victory over Soviet totalitarianism. The Iron Curtain's collapse was nothing less than an epochal achievement made possible by the cooperation between nine American presidents and congressional leaders from both parties. A continuity of policy and purpose was woven seamlessly through every presidency from Harry Truman's to George H. W. Bush's. That singular focus occasionally led to disastrous outcomes: the tragic legacy of Vietnam is shared by John F. Kennedy, Lyndon Johnson, and Richard Nixon. But Cold War diplomats like George Kennan, Henry Kissinger, Zbigniew Brzezinski, and Brent Scowcroft served both Republican and Democratic presidents alike, and their central focus across five decades—a commitment to toppling an "evil empire" responsible for the deaths of millions of its own citizens—ultimately proved to be a moral campaign that required great skill and endless fortitude.

Ronald Reagan and George H. W. Bush may have expedited and managed the Cold War's end, but it was Harry Truman whom history called upon to engage the

Soviet Union in a struggle for world supremacy during a time when most Americans were war-weary and exhausted by their role in resolving the endless tragedies of twentieth-century Europe. Truman knew ignoring Stalin's expansion across Central Europe was as unwise as the appeasing of Hitler's advances across that same continent a decade earlier; and yet, like his predecessor Franklin Roosevelt in the lead-up to World War II, Truman faced resistance from Republicans in Congress and progressives in his own party. Worse yet, many of Truman's own allies considered the new president ill equipped to confront Soviet expansionism and lead America through the postwar crises it would soon confront.

The journalist and historian Herbert Agar called Truman a "strange little man," and others shared that harsh assessment when the former Missouri senator assumed the presidency after Franklin D. Roosevelt's death in April 1945. Just twenty years earlier, the failed businessman had been defeated in his bid for reelection as county judge back home in Independence. The machine backing Truman thought so little of him that party bosses only allowed him to campaign for the United States Senate in 1934 after their first four choices refused to run. Even after his surprising win, the *New York Times* dismissed the incoming senator as a "rube." Truman would spend the next six years absorbing a steady stream of abuse while championing FDR's New Deal agenda, but his unremitting loyalty was not returned by the Democratic president, who refused to campaign for Truman's reelection bid or even endorse him. The *St. Louis Dispatch* dismissed Truman as a "dead cock in the pit" in that reelection battle against a popular Missouri governor. But Harry Truman would run and win as an underdog again, and once more shock Washington insiders who shared the *Dispatch*'s grim take on his chances. That would be far from the last time the five-feet-nine-inch commoner with an explosive temper and poor eyesight exceeded all the low expectations placed upon his narrow shoulders.

Four years after refusing to endorse Truman's reelection bid, a dying Roosevelt glumly selected the unrefined senator as his vice presidential pick in 1944. An FDR aide later admitted, "We chose Truman because all of us were tired." Roosevelt's own chief of staff, Admiral William Leahy, brusquely asked his boss during those deliberations, "Who the hell is Harry Truman?" The media's reaction was no less caustic. *Time* magazine responded to Truman's nomination by describing him as "the mousy little man from Missouri." Others mocked FDR's VP choice as "the Second Missouri Compromise."

Even after Harry S. Truman's elevation to the White House, the slights continued. "To err is Truman" was a jab Republican and Democrats alike used to target a politician who, at times, seemed hopelessly over-matched by the challenges confronting his presidency. My own family, who suffered through the darkest days of the Great Depression in rural Georgia, survived those desperate times because of Roosevelt's New Deal; a picture of FDR hung on my grandmother's wall until she died at the age of ninety-three, a half a century after Roosevelt's own death. And yet my mom remembered with a laugh her poor, rural family mock-ing the lack of abilities exhibited by fellow Baptist and Democrat Harry S. Truman. Like Washington elites and press barons in New York City, the Clarks of Dal-ton, Georgia, saw Truman as a country bumpkin unfit to lead the country, much less carry through on Roo-sevelt's legacy.

When President Truman sought reelection in 1948, some Democrats even ridiculed their eventual nominee with signs declaring "I'm Just Mild About Harry." But soon Truman would turn those mocking sneers into the fevered cries of "Give 'em Hell, Harry!" That bruising reelection campaign ended with the *Chicago Tribune*'s infamous headline "Dewey Defeats Truman" being printed as election-night experts declared the president a beaten man. Truman brushed aside those predictions and, instead of worrying about his political fate, went to bed. He would awaken the next morning to learn that once again, the failed haberdasher had proven his critics wrong, but this time in the most remarkable of ways; Harry S. Truman had somehow managed to pull off the greatest upset in the history of American politics. This modest man from Missouri had once again accomplished what he had done before as vice president, United States senator, and county judge: wildly exceeding the grim expectations laid upon him.

All too predictably, Truman's reelection victory would soon be followed by four more years of bruising attacks from both the left and right. And after eight grueling years in the Oval Office, Harry Truman left Washington with the lowest approval ratings ever recorded for a president. Though few of his contemporaries appreciated Truman's long list of history-bending achievements, the man who spent his life haunted by business failures, personal debt, and withering political criticism would return home to Independence, Missouri, as the most consequential foreign policy president of the past seventy-five years. Only his predecessor, Franklin Roosevelt, compares with Truman as commander in chief for shaping world events over the twentieth century.

The twelve presidents who followed the man from Independence inherited an international stage shaped by Truman's policies. Joseph Stalin's plans for expansion into Western Europe were undermined by containment, the Marshall Plan, the formation of NATO, the Berlin Airlift, and yes, the Truman Doctrine. The thirty-third president's decision to draft the disgraced former president Herbert Hoover to lead America's response to the European refugee crisis following World War II led to the two presidents saving more lives "than any two players on the world stage." As Nancy Gibbs and Michael Duffy wrote in *The Presidents Club*, Truman and Hoover's heroic work in feeding millions left starving across Europe after the war carried with it both a humanitarian and a strategic purpose. "Bare subsistence meant hunger, and hunger meant communism," Hoover told Truman. The Democratic president agreed and used the powers of his office to bring an end to that epic suffering that was crippling Europe. Taken together, the words, actions, policies, and legislation of Truman saved millions of lives across Europe and assured that millions liberated from Hitler by Allied forces would remain free in the postwar world. Even the lands enslaved by Soviet communism after the war would eventually be liberated because alliances and international structures left in place by Truman, in time, would lead to the Berlin Wall's collapse in 1989 and the Soviet Union's demise.

Winston Churchill declared of Truman that he, "more than any other man . . . saved civilization." And the legendary secretary of state Dean Acheson rightly praised this not-so-common man from Missouri for bringing about a "complete revolution in foreign policy." He had countered almost two centuries of Americans' deep suspicion over US involvement in the affairs of other nations and had somehow forged a path that made it possible for America and its NATO allies to save Europe from the scourge of Stalinism. But America's isolationist instinct had held sway from the Republic's first days, and is again leading to America's retreat across the globe.

In its time, it was the "Truman Doctrine" that finally brought one hundred and fifty years of American isolationism to an end.

Despite the sickening human and economic losses it suffered, the British Empire had still grown in the wake of the First World War. But while the United Kingdom held the future of Western civilization on its shoulders during the Battle of Britain in 1940 and 1941, its empire began to rapidly disintegrate after that war's end. The heroic and lonely stand taken by the British under Winston Churchill against Hitler's war machine wreathed that nation in everlasting glory, but exhausted its resources and its people. Their "finest hour" was followed by a dizzying decline, and their defense of freedom had broken Britain's back to such an extent that it could no longer play the role of a history-shaping world power. Its sprawling possessions, on which the sun had famously never set, were now unaffordable indulgences. Instead, the United States would bear the burden formerly borne by Great Britain as a guarantor of world peace. Pax Britannica would now become Pax Americana, and those countries not formally a part of the empire, but still under British protection, would also be left to their own devices.

Greece was an example of the latter. An ancient land dotted by the landmarks of classical civilization and surrounded by the sparkling waters of the Mediterranean, Greece was of prime strategic importance in the great power struggles of the nineteenth century. For centuries a part of the Ottoman Empire, Greece won its independence with British help in 1832. Its proximity to the Black Sea made its control key to Russian access to the Mediterranean; the British both maintained a military presence and provided economic support.

Greece would remain a battleground in the standoff between the West and Soviet communism. The Germans invaded during World War II, and even before V-E Day in 1945, the country was ravaged by a civil war between government forces and communist insurgents.

Under the banner of the Truman Doctrine, the United States pledged to "support free peoples who are resisting attempted subjugation by armed minorities or by outside pressures." This was a profound transformation of America's conception of itself and its role in the world. Under the precepts of Washington's Farewell Address, the United States would stand apart from international intrigue as an example for other nations to follow. But the principles of the Truman Doctrine transformed the US into an active participant in the political affairs of other endangered nations. A principle and worldview to which America had largely adhered since the presidency of George Washington was giving way to the realities of a shrinking world. The consequences for the United States and that world were vast.

America's increased engagement with foreign crises and global challenges was anything but inevitable. It came about only because of the skill and determined leadership of a group of statesmen led by Truman, who believed history required them to create a new and more just world in America's image.

But it was on Capitol Hill that the fate of Truman's postwar vision rested. Members of the House and Senate, and especially the latter, would ultimately determine whether America would finally assume leadership of the free world. The Senate, long a bastion of isolationism, in which Washington's Farewell Address has been read from the dais every year since 1862, was still alive with the spirit of Henry Cabot Lodge and those who had dashed President Woodrow Wilson's dreams of a League of Nations following the First World War. But Truman was determined things would be different following World War II. This would happen, in large part, because a formerly dedicated isolationist would undergo one of the most far-reaching and consequential political conversions in American political history. Senator Arthur Vandenberg of Michigan, chairman of the Foreign Relations Committee, would respond to Truman's call and set about convincing his reluctant colleagues that America could no longer concern itself merely with its own affairs.

Then, as now, the government of the United States was restrained by Madisonian checks and balances meant to frustrate ambitious would-be autocrats while producing measured legislation. The separation of powers, carefully designed by the Founders to foster deliberation and avoid precipitate action, made it impossible to produce sweeping policy shifts without coordination and compromise between the executive and legislative branches. The legislative process has long been complex and arcane, with most bills destined for oblivion; the Senate has been especially notorious for its ponderousness. Each member has looked in the mirror to see a future president looking back, and this self-regard convinces too many that their every lengthy oration is necessary for the enlightenment of the republic. One senator can still hold up the process indefinitely until their particular concerns are addressed. The House could in theory act more swiftly, but its larger size means that even more personalities are involved. The reconciliation of differences between House and Senate bills add yet another step to this constitutioinally measured but politically tortuous process.

An added obstacle—one that was deplored by the Founders but that emerged practically before the ink of the Constitution was dry—is the scourge of partisan conflict. Just as Washington's cabinet—and the Congress with which he contended—were divided between partisans of Great Britain and revolutionary France, twentieth-century Democrats and Republicans held profoundly different views about the role of government and the conduct of world affairs. The Republicans had evolved from their more pro-government origins into a party that believed in limiting the growth of the state. At least for the first half of the twentieth century, the party was also the natural home of isolationism. Suspicious of overseas engagements while begrudging their costs, most Republicans wished to restrain the international ambitions of Democratic presidents from Wilson to Roosevelt to Truman. In addition, Republicans were frustrated by the exceptionally long period of Democratic rule in Washington, resulting from Roosevelt's unprecedented four election victories. Fortunately for Harry Truman, there were those who understood that the then traditional Republican approach to foreign policy was ill suited for an age when Stalin and the Soviets looked to expand their empire into Western Europe. Still, partisanship remained an enduring element of American politics, and the constant struggle for electoral advantage would influence—if not determine— the intense debate that followed.

The conditions that compelled the president to announce the Truman Doctrine were acute, and the roiling international scene would not afford Truman and his cabinet the luxury of time. Could the government, subject as it was to the desires of a traditionally isolationist population of one hundred million, respond quickly enough to face down the onrushing events that threatened to overtake America and its European allies? Would the United States move away from the Farewell Address, written by its most revered statesman when the country was still young, to a hastily conceived doctrine proclaimed by a man who not long before had been a failed haberdasher? How American government answered these questions during Harry Truman's presidency would determine the future of Western civilization.

Chapter 1
America Alone

D ean Acheson's office was a mess.

Tall, handsome, and impeccably dressed, with a neatly clipped mustache, Acheson did not look like a man who thrived amid chaos. But on this cold and clear February morning, there was a good reason for the disorder surrounding the undersecretary of state. Along with the rest of his department, Acheson was preparing to leave his ornate quarters. The Department of State had occupied the south wing of the vast State, War, and Navy Building (today called the Eisenhower Executive Office Building), a French Second Empire–style edifice next to the White House, for more than sixty years, but was about to decamp to new headquarters in nearby Foggy Bottom. Mark Twain had called it "the ugliest building in America," and President Truman considered it a "monstrosity," but the old bureaucratic headquarters had provided a rococo counterpoint to the neoclassical mansion next door, and its European pretensions suggested the burgeoning role the United States government wished to play on the world stage. For most of State's occupancy of the old building, that aspiration remained unfulfilled.

Whatever its architectural merits, the State Department was an uncomfortable place to work. Acheson, who occupied lofty rooms connected to the secretary's office by a narrow corridor, would later recall that he and his colleagues "stifled under the full blast of the summer sun aided by its reflection from the roof of the portico just under the corridor window and unabated by any such newfangled contrivance as air conditioning. We stifled all winter, too, through equal inability to control the government's heating system."

The War and Navy departments had outgrown their allotted space and moved elsewhere, and now State would soon follow, allowing the "monstrosity" to become a part of the White House complex, housing the Executive Office of the president. The transfer to Foggy Bottom would provide a vast increase in space, and would symbolize a degree of independence from the watchful gaze of the president and his aides. Countless records, a veritable archive of American diplomacy, had to be packed and transported.

But the upheaval attending the move was nothing compared to the changes set in motion by the delivery that day—Friday, February 21, 1947—of two "blue papers," as formal diplomatic notes from the British Foreign Office were known. The British ambassador, Lord Inverchapel, had been instructed to deliver the notes directly to the secretary of state, General George Marshall. But Marshall, who had only recently returned from a lengthy diplomatic mission to China to take up the post of secretary, was away in Princeton to deliver a speech. There were diplomatic niceties to be observed; the ambassador could not be seen to deliver the notes to a lowly clerk. Thus, H. M. Sichel, the first secretary of the British embassy, was dispatched from the big Sir Edward Lutyens–designed British embassy on Massachusetts Avenue to the old gray building on Seventeenth Street and Pennsylvania Avenue.

With a practiced indifference that belied their stunning historic impact, Sichel delivered the notes to Loy Henderson, a senior State Department official with responsibility for the Near East and Africa. While their contents did not come as a complete surprise, as the British and American governments had for months been discussing the strategic importance of Greece and Turkey, the impact of their delivery on international affairs was seismic.

Henderson immediately conferred with his colleague John D. Hickerson, deputy director of the Office of European Affairs. Both immediately grasped the far-reaching implications of the British diplomatic message, and the new responsibilities that would soon be thrust upon the United States. They wasted no time in delivering the notes to the undersecretary.

Acheson later described them as "shockers," and knew in the instant that he read them that the world had changed forever.

The first of the two was the biggest shocker. It began innocuously enough: "His Majesty's Government are giving most earnest and anxious consideration to the important problem that on strategic and political grounds Greece and Turkey should not be allowed to fall under Soviet influence." But as Acheson read further, the full import of the note became clear. After a summary of previous discussions between the British and American governments on the topic, and an accounting of British support for the two troubled nations up until that time, the otherwise bland dispatch climaxed with the following stark admission: "His Majesty's Government have already strained their resources to the utmost to help Greece and have granted, or undertaken to grant, assistance up to 31st March, 1947 to the amount of £40 million. The United States Government will readily understand that His Majesty's Government, in view of their own situation, find it impossible to grant further financial assistance to Greece." There was no time to lose; the deadline was less than forty days away, and the note concluded with a request that "the United States government will indicate their position at the earliest possible moment." With much of Greece on the verge of starvation and the royalist government under constant rebel assault, a failure to act by the United States would be catastrophic.

The second note concerned Turkey. The British and American governments had previously committed to share the burden of military and economic aid to enable it to resist outside influence. Weakened by war and the still fairly recent loss of its once-large empire, Turkey was vulnerable to the voracious appetite of the Soviets, who became focused on the strategically vital Dardanelles, off Turkey's Gallipoli peninsula. But the British were now warning their American allies that they would no longer be able to honor their previous offer of economic aid, and any future involvement would be limited to a modest military presence. The guarantee of a stable and free Turkey would now be placed squarely on Truman's shoulders.

Lord Inverchapel was the diplomatic representative of an empire both bankrupted and broken by two world wars. From 1939 to 1945, Britain had extended itself militarily and economically in nearly every corner of the world, and its citizens had paid a heavy price. Sixty thousand civilians had been killed in merciless German bombing raids during the Battle of Britain, and many more citizens lost their homes. Wartime rationing had made luxuries out of the most basic necessities of life. And in the aftermath of that epic suffering, Britain faced one of its harshest winters in history just months after the war's end. Massive snowdrifts buried vast sections of the country, paralyzing transport and further damaging Britain's economy.

Thus did the mighty British Empire, which had ruled the waves and much of the world for centuries, pass the baton of world leadership to its former colonies. In a speech at West Point years later, Acheson would famously observe that Britain had lost an empire and not yet found a role; he might have mentioned that he had been a witness to that demise. Just as Churchill stood alone against Hitler's war machine in 1940, it would now be Harry Truman's government standing alone against Stalin's designs on Western Europe seven years later.

Acheson swung into immediate action, ordering his staff to gather relevant personnel from the European and Near Eastern divisions and prepare reports on Greece's and Turkey's economic and military needs. There would be no rest for the State Department's leadership that weekend. But Acheson's team knew what was expected of them, and appreciated his managerial approach. As one aide recalled:

Meeting with members of his own staff, Acheson never stated an opinion or conclusion until everyone present had an opportunity to give his own ideas about the subject and suggest a remedy. By questions he stimulated others to talk, while he listened and took occasional notes. When every aspect of the matter had been carefully and fully considered, he would summarize what he had heard, point out conflicts in points of view, attempt to reconcile them, introduce facts and reasoning that might not have appeared, and finally suggest a solution. It was as though he were aware that his own logic and facility of expression might, if brought into play too early, intimidate full expression. Doubts usually fell away at the close of Acheson's meetings and ended in agreement, each person present feeling that his view had made a major contribution.

Only then did the undersecretary of state pick up the telephone. It was time to brief the president, a task he had performed many times before. Truman and Acheson had established a cordial and comfortable relationship through their regular meetings in the Oval Office. But with this call, he was alerting the commander in chief that a new era in American foreign policy—and world history—had begun. The president had a keen sense of the past; he once said to his aides, "If a man is acquainted with what other people have experienced at this desk, it will be easier for him to go through a similar experience. It is ignorance that causes most mistakes." Having already made more momentous decisions in two years than most presidents do over two terms in office, Truman also had a wealth of personal experience to guide him. The United States was fortunate at that critical juncture to have a chief executive whose outlook was influenced by the example of his most successful predecessors.

In characteristically brisk and businesslike fashion, Truman endorsed Acheson's initial actions and ordered that a report be delivered to him and Secretary of State George Marshall by Monday. The story of their fruitful diplomatic collaboration, which would culminate in Acheson's appointment as secretary of state in 1949 (a post he would hold for the entirety of Truman's second term), was about to begin one of its most critical chapters.

Henderson and Hickerson wasted no time. Both were seasoned diplomats with decades of experience who had held numerous postings across the world. As seasoned observers of the international scene, they recognized the momentous shift that was under way, but the British memos had confirmed their fears while putting them in a position to finally begin implementing the ideas and policies they had both long nurtured.

That evening, Acheson's team gathered their staff together to analyze the challenges facing US policy makers and prepare a memorandum for Marshall's review.

The United States was blessed at that moment with a formidable array of diplomatic talent, but perhaps the most intellectually gifted of those meeting that evening to sketch out the future of America's foreign policy was George Kennan. Forty-three years old, a native of Milwaukee, Wisconsin, Kennan had steeped himself in international affairs since childhood, moving to Germany and learning the language as a young boy. After graduation from Princeton, he joined the Foreign Service and had postings throughout Europe before deciding to specialize in Russian affairs. His linguistic abilities and intellectual gifts granted him an understanding of the arcane workings of the Soviet Union beyond that of most of his contemporaries. Kennan was horrified by the brutality and oppression that Stalin inflicted upon the Russian people and convinced him that the dictator harbored expansionist designs. While in his first Russian posting, Kennan found himself out of step with Ambassador Joseph Davies, who was a hapless diplomat somehow seduced by Stalin's rough charm. But several years later, as deputy chief of mission under Ambassador Averell Harriman, Kennan's warnings about Stalin's Soviet Union found a more receptive audience.

It was with his "Long Telegram" of February 1946—a foundational document in American Cold War foreign policy—through which George Kennan established himself as an invaluable player in Washington. This lengthy diplomatic cable to his superiors in the Truman administration warned:

At bottom of Kremlin's neurotic view of world affairs is traditional and instinctive Russian sense of insecurity. . . . For Russian rulers have invariably sensed that their rule was relatively archaic in form, fragile and artificial in its psychological foundation, unable to stand comparison or contact with political systems of Western countries. . . . They have learned to seek security only in patient but deadly struggle for total destruction of rival power, never in compacts and compromises with it.

It was no coincidence that Marxism, which had smoldered ineffectively for half a century in Western Europe, caught hold and blazed for first time in Russia. Only in this land which had never known a friendly neighbor or indeed any tolerant equilibrium of separate powers, either internal or international, could thrive a doctrine which viewed economic conflicts of society as insoluble by peaceful means.

With Kennan in the chair, there was little doubt that the meeting would swiftly produce a set of vigorous policy recommendations. No notes on the discussion survive, but history records that none around the table that night had any reservations about what the United States had to do next. This gathering of like minds was determined to see America take its rightful place as the leader of the free world. Kennan and the group agreed that a draft memorandum should be immediately prepared that evening and further reviewed over the weekend.

On Saturday, further meetings were held that included senior military officials whose expertise and blessing would be vital if the initiative was to succeed. Though weary from four year of war with Germany and Japan, most military leaders had long cast a skeptical eye toward Moscow. The memorandum was revised and redrafted as officials raced against the clock to provide the most comprehensive case they could for saving Greece and Turkey.

Finally, it was done. In terse, uncompromising language, the memorandum warned that if the United States failed to come to the aid of those fragile nations, it "would lay these countries open to Russian domination," and Britain "may consider herself compelled to pursue policies of her own with regard to these countries." The consequences would be dire: "a widespread collapse of resistance to Soviet pressure throughout the Near and Middle East and large parts of western Europe not yet under Soviet domination." The memorandum urged a program of both military and economic aid, declaring that "half-way measures will not suffice and should not be attempted."

The memorandum recommended "further steps" should President Truman approve the proposal, including discussing the issue "privately and frankly . . . with appropriate members of the Congress," drafting the necessary legislation, and adopting measures "to acquaint the American public with the situation and with the need for action along the proposed lines."

Less than two years after the Allied victory in Europe, the United States was now on the verge of leading a fight against one former ally while the other sat helplessly on the sidelines. It was time for the American president to lead the world in its fight against communism. But would Harry Truman dare to engage in that historic struggle? And if he did, would Americans follow?

Chapter 2
Greek Fire

O n a cool spring night in 1946, thirty-three communist guerrillas descended from their hideout atop Mount Olympus, the highest mountain in Greece and the mythological home of the gods. Down its rocky sides they scrambled, ready to wreak havoc on the towns below. The communist insurgents' timing was not random: a general election was to be held the following day, and the guerrillas had pledged to overthrow the government in Athens through a campaign of violence and terror. Given the chance to participate in the democratic process, the communists chose violence instead despite the fact—at the same time—Italy's communist party was contending for elective office. But in Greece bullets were chosen instead of ballots, and disastrous results would follow.

The communist force ran down silent streets and attacked without mercy. Orange flames filled the night skies, and the stars over Greece were obscured by smoke, as guerrillas set fire to several buildings and shot anyone who got in their way. As had happened in Greece so many times over the past several years small villages paid the price for national political quarrels. This communist attack was just the latest chapter in the country's long bout with civil unrest.

This was the opening scene in the latest phase of the Greek Civil War, a bloody internecine conflict that had begun while the Second World War was still raging and would continue intermittenly for decades to come.

This Greek Civil War would prove to be much more than a mere regional struggle. By an accident of history and geography, it became a flash point between the world's most powerful nations, the United States and the Soviet Union. The streets once roamed by philosopher kings and the first architects of Western Civilization would soon become the first theater in the Cold War between the capitalist West and communist East.

The wild landscape of northern Greece was ideal terrain for guerrilla warfare. For hundreds of miles, along the borders of Albania, Macedonia, and Bulgaria, the region offered endless opportunities for ambush. The guerrillas could easily move across the borders, and on either side win new converts to their cause. Since the days of the Ottoman Empire, bandits had roamed freely, preying on travelers foolhardy enough to venture there. No occupying force—not the Turks, not the Germans—found it possible to police such wild and inhospitable land, and the communist insurgents were proving themselves to be a tough and resilient force fired by ideological zeal and focused on the overthrow of the existing political order. Of course, what they proposed in terms of replacement would prove to be far more oppressive than even the current chaotic situation.

The Communist Party of Greece (KKE) had one crucial advantage over its royalist and nationalist political rivals: like the communists in France, it had been one of the prime enablers of resistance to the German occupation. It sponsored the powerful National Liberation Front (EAM), the most important organization devoted to fighting the Germans, cleverly camouflaging its communist goals with nationalist rhetoric. The EAM in turn created a military force known as the People's Liberation Army. While there were other noncommunist resistance groups, none were as resolute in their anti-Nazi activities. Even during the Second World War, the EAM moved ruthlessly to crush competing groups engaged in the struggle against the

Germans. So merciless were its tactics that some resistance fighters even decided to throw in their lot with the Nazis.

But the communists had more than just terror and violence as a tactical advantage. The ravages of war and the resulting privation and despair increased the appeal of the EAM's ideology. The Nazi occupiers had had little regard for the well-being of the Greek people. *Reichsmarschall* Hermann Göring had exhorted military commanders in Greece and elsewhere to "get hold of as much as you can so that the German people can live. . . . I could not care less when you say that people under your administration are dying of hunger. Let them perish so long as no German starves." While the EAM led violent campaigns, it also fed the population, founded schools, and provided other services that the government in Athens could no longer offer. This combination of ruthless force and public philanthropy made the communists a force to be reckoned with, and in the wild and remote mountains of northern Greece, they were quite literally a law unto themselves. One of the commanders declared, "This is a revolution. And things have to be done—even if a few innocents are killed, it won't matter in the long run."

While the Germans had invaded Greece in April 1941, Italian troops had already invaded Albania on

Good Friday, 1939. The campaign lasted only a few days, and the Albanian king Zog and his family fled south to Greece. Eager to follow the example of his ally Adolf Hitler, and hungry for more territory, Italian dictator Benito Mussolini resolved to take Greece as well. In meetings with his ministers, Il Duce brooked no dissent, and despite having made no detailed plans for the invasion, he sent his Albanian-based forces into Greece on October 28, 1940.

But Mussolini's scattershot plan proved to be no blitzkrieg. The Italian army lacked the dash and deadly efficiency of the Germans, and in the mountains of northern Greece it met tenacious resistance from the Greek army. Rather than taking new territory, the Italians found themselves forced back into Albania little more than two months after the campaign began. The Greek National Army had proven to be resolute under the most dire of conditions.

Mussolini's failed invasion made a national hero of Greece's reviled dictator, General Ioannis Metaxas, who had seized power in 1936 and outlawed all political parties. When the Italian minister to Greece delivered formal notice of the invasion to Metaxas, the Greek leader famously (if perhaps apocryphally) pounded his desk and shouted, "*Ochi!*" ("No!") This soon became a rallying cry for the Greek people.

However impressive its performance, the Greek army remained badly outnumbered and low on supplies. The government called upon the United Kingdom for help, and British troops arrived there in March 1941. "I felt more like a bridegroom than a soldier with my truck decorated with sprigs of peach blossom and my buttonhole with violets," said one British officer of the lavish reception granted him and his comrades by the civilian population.

The Germans soon ended that short Greek honeymoon. Hitler, frustrated by his flailing fascist ally but unwilling to risk an Axis reversal before his planned invasion of the Soviet Union, sent an invading force to Greece through Bulgaria on April 6, the ferocity of which was undiminished by the fact that the Germans attacked Yugoslavia at the same time. Other than to reverse Mussolini's embarrassing performance, Hitler had no interest in a Greek campaign and forbade the Luftwaffe from bombing Athens. But no such restraint was applied to the mountains of northern Greece; the air was soon thick with German Stukas.

The Greek army fared even worse against the German onslaught than had French forces the year before;

Greece swiftly capitulated. The British began a ragged and undignified retreat on April 20. Within three weeks, the Germans entered Athens in triumph and soon after occupied Crete.

The Acropolis, an ancient hilltop citadel designed by Pericles, was defaced by a swastika flying overhead, and the German occupiers proved to be so ruthless in their policy of plunder that one hundred thousand Greeks died in the coming winter. The king of Greece, George II, fled to Cairo with his government. Like so many displaced royals during the war, he ruled his country from abroad under British supervision. That supervision was tinged with condescension, notwithstanding the familial connections between the Greek royal family and the British aristocracy. (Not long after the war, in yet another example of this intermingling, George II's first cousin, Philip, prince of Greece and Denmark, would wed Princess Elizabeth, the future queen of England, and come to be known as the Duke of Edinburgh.) Winston Churchill's intent was to restore the Greek king to his throne once the war was won.

American military strategists such as army chief of staff George C. Marshall were keen on a cross-channel invasion as early as 1942, but Churchill and Marshall's British counterpart, General Alan Brooke, chief of the Imperial General Staff, wished to postpone a frontal assault until the Germans could be harried at the margins and Allied troops had gained more experience. This led to strained arguments at Allied conferences, but Churchill's view eventually prevailed. Churchill had often been accused of fighting peripheral battles rather than taking on the enemy directly. But in the case of Greece, it is arguable that the British presence during the German invasion slowed the pace of Hitler's troops just enough to delay their surprise invasion of Russia until June 22, late enough that the Russian winter later helped contribute to the disastrous German defeat there.

Churchill's relevance to Greece's fate didn't end with that strategic decision. His tactics deployed in a brief meeting with Joseph Stalin was to have enormous implications for the postwar world. On a visit to Moscow in October 1944, Churchill met with Stalin to stress the vital importance of Greece to Britain. This was an example of the individual diplomatic course taken by both Roosevelt and Churchill during the war; they may have been the closest of allies but their strategic priorities were far from being in alignment. Churchill had declared in 1942 that he had "not become the King's first minister in order to preside over the liquidation of the British Empire," and in his discussions with Stalin he showed his willingness to pay a price to avoid that unhappy fate. Acknowledging that "the Americans might be shocked" by their conversation, Churchill pressed Stalin for a guarantee that he would not interfere in any Greek postwar settlement, in exchange for giving the Soviets a free hand in Romania. Notes of the meeting record that Churchill "then produced what he called a 'naughty document' showing a list of Balkan countries and the proportion of interest in them of the Great Powers." As Churchill later recalled, he wrote on a "half sheet of paper" his proposal:

Roumania
    Russia   90%
    The others   10%
Greece
    Great Britain (in accord with USA)   90%
    Russia   10%
Yugoslavia   50–50%
Hungary   50–50%
Bulgaria
    Russia   75%
    The others   25%

Churchill remembered later that after he gave the slip of paper to Stalin,

There was a slight pause. Then he took his blue pencil and made a large tick upon it, and passed it back to us. . . .

After this there was a long silence. The pencilled paper lay in the centre of the table. At length I said, "Might it not be thought rather cynical if it seemed we had disposed of these issues, so fateful to millions of people, in such an offhand manner? Let us burn the paper."

"No, you keep it," said Stalin.

The "Man of Steel" could afford to appear magnanimous (if magnanimity meant agreeing to be the dictator of millions). He knew as well as anyone how precarious were Britain's finances, and how tenuous was her hold on various dominions. Therefore he honored this squalid but necessary bargain with Churchill, knowing that he could strike later once the British government was no longer capable of enforcing the deal. And besides, the communist guerrillas of Greece were more than capable of maintaining their campaign against the government without direct Soviet help. Guns and terror proved to be cheap commodities for the communists.

It was important to Britain that Greece remain democratic, a fact further underscored by Churchill's visit to Athens in December 1944. The war was still ongoing in both Europe and the Pacific, but the Germans had withdrawn from Greece in October, leaving its citizens to pick up the pieces amid a blasted political and physical landscape. The extreme ideological polarization caused by the occupation and the nature of the resistance left the government heavily dominated by right-wing forces as brutal as many of the communists they were now fighting.

Skirmishes had broken out between the British-backed government and communist insurgents who had been a part of the anti-German resistance. Churchill left his family to celebrate Christmas without him, and he and the foreign secretary, Anthony Eden, flew to the Greek capital on a mission to prevent a communist victory.

The prime minister convened a meeting of the leaders of both sides in the conflict, urging them to come to a compromise while reminding them that British troops would intervene if necessary. The results were mixed: fighting continued after his departure, but all sides accepted the appointment of Archbishop Damaskinos as regent, a holding action until the Greek people decided whether they wished to restore the exiled King George II to the throne.

The British wanted a friendly and stable Greece, but

Churchill was clear that the political details were up to the Greeks, who would be "at liberty to choose by free elections the sort of constitution and government they desired." This was not a right they would be granted were the communists to prevail.

Perhaps because of their disdain for democracy, the communists decided not to contest the March 1946 elections but instead to resort to terror and violence to discredit the rightist government. Their strategy backfired and had the effect of strengthening the right while further marginalizing the communists. Left unconstrained, the right's hold over Greek politics led to the government arresting and executing large numbers of communists, while virtually ignoring the wartime collaborators still in their midst. This brutal display would create complications for Truman the following year as he tried to convince Congress that supporting the Greek government would be in the national interest. Their opponents would be able to claim, with some justification, that by backing the Athens government the United States would be propping up a regime every bit as brutal as any communist regime that might supplant it.

In October 1946, the ragged bands of communist guerrillas who had escaped government reprisals were united to form the Democratic Army of Greece (DAG), under the leadership of Markos Vafiadis. The DAG combined traditional guerrilla tactics with slightly more conventional command structures, and the roving bands of the earlier phase of the conflict increased in size by a factor of ten.

The DAG used training camps in Yugoslavia to prepare their troops for battle and provide constant indoctrination to fire their ideological zeal. This foreign interference made it even more difficult for the Greek government to squash the insurgency, for the guerrillas could retreat across the border much faster than the official Greek army could infiltrate the mountains.

The government looked to the United Nations for help, with results that would help propel the events described later in this book. The UN, then only in the second year of its existence, formed a commission to investigate the situation. Predictably, the commission produced a lengthy report that promptly began gathering dust in UN archives. Russian representation on the commission—unavoidable because Russia was a member of the Security Council—did not prevent the commission from acknowledging that Greece's neighbors to the north were acting aggressively toward the sovereign nation. But the Russians and their Polish puppets ensured that nothing would be done to stop their activities.

No longer simply a relic of the classical past, dotted with splendid ruins and surrounded by sparkling blue seas, Greece was now ground zero in the contemporary clash of empires. The Cold War was about to begin in Athens.

The British Labour government that had swept Winston Churchill from power in the stunning general election of 1945 was determined to decolonize the vast British Empire, and within two years had granted independence to Jordan and India, long considered the jewel in the British crown. But even Labour, at the urging of Foreign Secretary Ernest Bevin, Prime Minister Clement Attlee was determined to keep Greece within its orbit.

Britain was now faced with a difficult choice. Then as now, it was heavily dependent on the Middle East for oil, and Greece was the strategic key to that vital region. The British government was run by left-wing leaders determined to implement a socialist program that had been overwhelmingly endorsed by the electorate, but who had no sympathy for the ideological, expansionist aims of the Soviet Union.

But the economic realities facing an exhausted postwar Britain had made its path clear and unavoidable. Attlee knew he had no choice but to back away from long-standing commitments and instead tend to Britain's own domestic affairs. The British government could now do little more than hope its wartime ally across the Atlantic would embrace a newly expansive and unprecedented role throughout the world. The former colonies, which had won their freedom from the British Empire 170 years earlier, would now be asked to fulfill their former masters' global obligations. Legend has it that after their defeat at Yorktown, the British fifes and drums played "The World Turned Upside Down." Clement Attlee and his cabinet might have had the same tune on their minds as the British once again retreated to America's benefit.

The signs of that British retreat had long been apparent, and American diplomats in Moscow, Athens, and Istanbul had been increasingly concerned about the dangers of Soviet influence in Greece and Turkey. General Walter Bedell Smith, the American ambassador to Russia, had warned: "Turkey has little hope of independent survival unless it is assured of solid long term American and British support."

America was now required to take over where Britain left off. In a preview of things to come, the United States—through its Export-Import Bank—granted $25 million in aid to Greece at the beginning of 1946.

America's ambassador to Greece, Lincoln MacVeagh,

sent a steady stream of cables to the State Department in the winter of 1947 describing the grim situation. His cable of February 20 arrived on the eve of Britain's dramatic announcement to the Americans that it could no longer provide for Greece's security. In his cable, the ambassador warned that the situation was "critical," and continued: "Impossible to say how soon collapse may be anticipated, but we believe that to regard it as anything but imminent would be highly unsafe." The economic and military crises were challenging enough, but "deteriorating morale" due in part to "exploitation by international Communists" made the situation even more acute.

General George C. Marshall, who had been installed as secretary of state on January 21, summed up the situation in a cable to the ambassador: "Many former adherents of liberal and center parties, alarmed at presence of communists or condonement of communism, seem to have gravitated towards extreme right while others shocked at reactionary attitude of rightists have gone over to groups controlled or contaminated by communists." It was now left to Truman, Marshall, and the United States Congress to contain the rising tide of political chaos that was now overtaking Greece and threatening to sweep across the rest of Europe.

Chapter 3
Affairs of State

Henderson delivered the final memorandum to Acheson's house in Georgetown on Sunday evening. The neighborhood of row houses and cobblestone streets was the oldest in the city; it had been a Maryland village before the state ceded part of its territory to create the District of Columbia in 1791. Georgetown in Acheson's time was not yet the wealthy enclave it would later become, but many of the nation's political and diplomatic elite already called it home. Columnist Joseph Alsop lived a few blocks away, as did Philip and Katherine Graham, both future publishers of the *Washington Post*. Other historic figures would later join the "Georgetown Set," including Senator and Mrs. John F. Kennedy.

The undersecretary reviewed the material and pronounced it satisfactory. Henderson and his boss both felt prepared for what would be a momentous week in Washington. "At that," Acheson later recalled, "we drank a martini or two toward the confusion of our enemies."

As he sipped his drink in the old brick house on P Street, Acheson must have savored the moment. He and his staff had moved with remarkable speed and unanimity, even with the stakes looming large before them. But who was this unruffled diplomat with patrician manners and exquisite tailoring? In another life he might have been a British ambassador, and he even spoke with a mid-Atlantic accent that infuriated his critics. His father had been born in England, and as a child young Dean was raised to celebrate the queen's birthday.

But Acheson's latent Anglophilia never clouded his strategic vision. He understood better than most that the United States was poised on the brink of a new era of global leadership, and that it was the only nation that could take up the burden laid down by a faltering British empire. He dedicated most of his life to fulfilling that vision, spurning the life of indolence and ease that his privileged background and elite education would have afforded him.

Dean Gooderham Acheson was born in Middletown,

Connecticut, in 1893, son of an Episcopal priest who would eventually become bishop of Connecticut. His father had first emigrated to Canada before arriving in the United States, and there married the Canadian daughter of a prosperous businessman. Acheson enjoyed a happy and stable childhood. He was exceptionally bright and viewed the world around him as a playground ordered for his pleasure.

With little effort, and less interest, he passed through Groton, the austere and forbidding boarding school recently founded for the offspring of the nation's elite. A free spirit, he chafed at the school's rigid structure and proved to be a poor student. His classmates relentlessly hazed him, and he once faced expulsion. Only his mother's intervention kept him in the school.

But Acheson survived his Groton ordeal, and after an unlikely but enjoyable summer spent working on railroad construction in Canada, he entered Yale as an undergraduate student. The New Haven college was far different in 1912 than today. The fact that he had been last in his class at Groton did not deter the admissions office. Yale academics proved to be just as boring to young Acheson as Groton's, but Dean led an active social life and began to develop the sense of style for which he would become known.

Then Harvard Law beckoned, and the once-freewheeling student found himself enthralled by Professor Felix Frankfurter, the future legendary Supreme Court justice. Frankfurter encouraged him to take law seriously as a field of study, and Acheson began to excel academically for the first time in his life. He later recalled of his intellectual awakening, "Not only did I become aware of this wonderful mechanism, the brain, but I became aware of an unlimited mass of material that was lying about the world waiting to be stuffed into the brain." Groton's worst student would graduate near the top of his class at Harvard Law.

After a brief naval career near the end of the First World War, he secured with Frankfurter's assistance a clerkship with Supreme Court justice Louis Brandeis in 1919. Continuing his meteoric ascent, he joined the prestigious Covington, Burling, and Rublee law firm (now Covington & Burling). Acheson gained valuable experience dealing with cases of international law, and built a comfortable life with homes in Georgetown and rural Maryland. The firm's close ties to the Democratic Party would inspire his next career move.

Government service soon followed. The newly inaugurated President Franklin Roosevelt appointed Acheson undersecretary of the Treasury before his fortieth birthday. He would soon find himself serving as acting secretary during the illness of Secretary William H. Woodin until a policy clash with the White House led him to resign months later.

When the war came, Acheson returned to government as assistant secretary of state in 1941. His portfolio was vast in scale: he managed the all-important Lend-Lease program, and success there led to his influence in the administration growing exponentially. That portfolio expanded in part because of a vacuum of power just above him. The secretary, Cordell Hull of Tennessee, was a consummate politician who conferred with the president regularly, but the Roosevelt administration's foreign policy was run almost exclusively out of the White House. Cautious and laconic to a fault, Hull rarely dared to challenge Roosevelt's dominance. (He also suffered from a speech impediment.) Hull's nearly twelve-year tenure as secretary was the longest in the department's history, but his longevity derived in part from Roosevelt ability to ignore him whenever he wished. Acheson himself was beckoned to the Oval Office whenever it suited FDR, who had little regard for precedent and procedure. Roosevelt's reliance on personal relationships may have helped forge alliances that helped win the war, but it also caused no end of confusion throughout the staid corridors of official Washington. Fortunately for Acheson, his experience with Truman's predecessor proved to be excellent preparation for the events to come.

By 1944 Hull was unwell and resigned his office. He was succeeded by Edward J. Stettinius Jr., whose tenure was brief, undistinguished, and blighted by his acquiescence to Soviet bluster at the ill-fated Yalta Conference. There a sickly Roosevelt succumbed to Stalin's blandishments, and acceded without protest to the dictator's dark plans for Eastern Europe. Churchill fought to stem the tide as much as he could, but Roosevelt haughtily dismissed the British prime minister and traded jokes with Stalin at Churchill's expense. An anguished Churchill said to his daughter that evening, "Tonight the sun goes down on more suffering than ever before in the world." History recorded the summit as a thoroughly discreditable display on the part of the American delegation. As historian Max Hastings has observed, "It may be true that the Western allies lacked the military power to prevent the Soviet rape of eastern Europe, but posterity is entitled to wish that Roosevelt had allowed himself to appear less indifferent to it."

Perhaps because of the shame attached to Yalta, Truman wasted little time in replacing Stettinius with his former Senate colleague James F. Byrnes, whom Truman had pledged to support for vice president at the 1944 convention that had so dramatically changed his life. Byrnes was no diplomat, but he was ambitious and determined, and served the new president well—for a time. Unlike his predecessor, the former South Carolina senator had no illusions about the Soviets and fierecely confronted them throughout his nineteen-month tenure. But Byrnes's ambition—and his belief that the man from Missouri was in the place that Byrnes himself should have been—got the better of him. He began to exceed his authority, and Acheson—now undersecretary—found his management style intolerable. Few in Washington protested when Byrnes was sent packing in January 1947 and replaced with General George C. Marshall, who had just returned from a long and grueling mission to China.

Marshall's appointment immediately ushered in a more efficient operation at State. A keen spotter of talent and effective delegator, the former general prized Acheson's abilities and devolved upon him vast authority over the sprawling department. As the nation's chief diplomat in an age when travel was still an ordeal, Marshall knew the State Department would be in competent hands during his frequent absences. Acheson responded to this grant of authority by venerating Marshall and ensuring that his every move met with the general's approval. This relationship was key to the speed and efficiency of State's response to the British notes, and also allowed Truman to operate knowing his diplomatic team was united in its advice.

One of Acheson's many advantages inside Washington's sprawling bureaucracy was the great trust placed in him by President Truman. After the Democrats suffered a crushing defeat in the midterm elections of 1946, losing both the House and Senate, Truman returned to Washington to find only one man waiting for him at Union Station: Dean Acheson. This display of loyalty, which came naturally to the undersecretary, made a profound impression on the president. Truman would never forget it, and Acheson would one day be greatly rewarded, in part, for that small gesture.

The "Domino Theory" first posited by Dwight Eisenhower in 1954 suggested that one country's fall to communism would cause the collapse of others surrounding it. Though discredited by America's future failures in Korea and Vietnam, the fear of Soviet expansion was even more pronounced in 1947. Acheson reasoned that "if France went communist, Italy and Greece were through; if Italy went communist, Greece was through; and if Greece went communist Turkey was in trouble; and if they all went communist Iran was in trouble."

Dean Acheson's decisiveness that day was in character. "Wise in the ways of Washington," as Clark Clifford described him, Acheson saw the crisis as an opportunity not only for America to assert firm leadership in a new world order, but also for the State Department to take the lead in making policy. The military had been in the lead for so long, having triumphed in the greatest war the world had ever known, that the diplomatic service felt often overshadowed. But what was called for in Greece and Turkey was economic aid and diplomatic finesse. Now it would be leaders inside the Pentagon who would find themselves sitting on the sidelines.

Chapter 4
Met at Armageddon

On Monday, Acheson returned to work and waited for further direction from the president and secretary of state. Both Truman and Marshall knew that whatever plan they devised could only check Stalin's advances into Greece and Turkey if it first gained support from Republicans and Democrats in Congress. The task would be diffficult. Over lunch at the Metropolitan Club, he fretted to a journalist: "The trouble is that this hits us too soon before we are ready for it. We are having a lot of trouble getting money out of Congress." Among his many other tasks at State had been congressional relations. Acheson had taken the measure of members of the House and Senate, and the Groton and Yale man found them wanting. He saw the politicians there as beholden to narrow and parochial interests, incapable of understanding seismic changes buffering world affairs any more than they could grasp the need for the United States to adopt an increasingly expansive role. The undersecretary's toughness and determination would help in the legislative struggle to come, but his patrician manners and unconcealed sense of superiority would predictably alienate many on the Hill.

During his recent trip to Princeton, Marshall had laid some of the rhetorical groundwork for the policy changes to come. On Saturday, February 22, as his subordinates scrambled to prepare the department's draft response to the British "blue papers," the secretary of state delivered a speech at the university's Alumni Day festivities in which he sounded the themes of world engagement:

Now that an immediate peril is not plainly visible, there is a natural tendency to relax and return to business as usual, pleasure as usual. . . . It is natural and necessary that there should be a relaxation of wartime tensions. But I feel that we are seriously failing in our attitude toward the international problems whose solution will largely determine our future. The public appears generally in the role of a spectator—interested, yes, but whose serious thinking is directed to local immediate matters. Spectators of life are not those who will retain their liberties nor are they likely to contribute to their country's security.

Marshall returned to the State Department on Monday the twenty-fourth and received a briefing from Acheson, as well as his memorandum on the crisis. Given the theme of his Princeton address, it is not surprising that Marshall proved to be a receptive audience when the British ambassador formally presented the messages the State Department had received on Friday. The secretary gravely responded that he would seek an answer from the president as soon as possible, and prepared to cross West Executive Avenue for a cabinet meeting with Truman.

Marshall arrived early, and asked to see the president alone in the Oval Office. He told Truman about what had transpired since the British notes were delivered on Friday, and assured him that a coordinated response was being prepared and would soon be ready for his review. Marshall also passed along a memo from Acheson in which the latter expressed his belief that the British were "wholly sincere in this matter and that the situation is as critical as they state. This puts up the most major decision with which we have been faced since the war."

Acheson led further discussions on the crisis with the secretary of the navy, James Forrestal; Secretary of War Robert Patterson; and members of the top brass of both services. They quickly determined that immediate action regarding Greece and Turkey was in the national security interests of the United States, and that the president should seek congressional support at once.

Having secured unanimity within the upper reaches of the State Department and the White House, Acheson on February 25 assembled a wider group of department officials to sound out their views. A few aired reservations, especially about a possible Soviet response—but most of those assembled supported the proposed policy and were determined to see it come to fruition.

Hours later, Acheson and his team put the final touches on a document entitled "Position and Recommendations of the Department of State Regarding Immediate Aid to Greece and Turkey." In bureaucratic prose that nonetheless crackled given the vast consequences of the policies proposed, the document warned that if the United States did not fill the void created by the British, then the recent victory over Nazi tyranny would be squandered, with grave consequences for the democratic West.

Now that State's position was finalized, Acheson needed to win the final approval of secretaries Forrestal and Patterson. Momentum would be lost if the administration was divided, and Acheson needed to go to the president with a plan backed by all the relevant players on his team. During their meeting on February 26, the two military chiefs expressed their support for the plan, but raised concerns. Forrestal and Patterson were rightly focused on several other hot spots around the world, and warned that Greece and Turkey were not the only countries in need of American aid. Korea and China were both in a precarious state, and other nations were sure to come hat in hand. But as Joseph M. Jones relates in his history of the period, the urgent situation in Greece made it necessary for other vulnerable countries to take a back seat for the moment. With luck, success in Greece and Turkey might increase the possibility of coming to the aid of other beleaguered countries in the future. The military leaders concurred.

Following their fruitful meeting with the secretaries of War and the Navy, Marshall and Acheson formally presented the recommendations to the president that afternoon. Truman was now ready to move.

Having his administration united and speaking with one voice on the question of aid to Greece and Turkey, it was time for the president to gain the support of Congress. This would require a speech—perhaps even a joint session—but first there was groundwork to be laid.

The challenge facing Harry Truman was made greater by the disastrous results of the 1946 midterm elections a few months before. Republicans had won resoundingly, gaining 57 seats in the House and 13 in the Senate, seizing control of both chambers for the first time since 1932, the dawn of FDR's Democratic coalition. Republicans had been elected on a familiar platform of tax cuts and small government, and were in no mood to pass new and expensive initiatives proposed by the Democratic president, especially if the money was being spent on foreign countries. Had the American people not sacrificed enough over four bloody years of war? Had the US Treasury not been squeezed to a breaking point throughout World War II? Was it not time for a return to normalcy when America's own needs could finally be put first?

Congressional Republicans certainly thought so. The president's first federal budget request after the election landed with a thud on Capitol Hill. Truman's proposed $37.5 billion budget was less than he had originally planned, but more than congressional Republicans were prepared to accept. The only question among his opponents was precisely how much to slash. (The spending cuts were not merely for the sake of reducing federal expenditures, but also to make possible a substantial tax cut. In those quaint times, American politicians still believed that revenues and expenditures should remain roughly equivalent.)

Despite howls of protest from the military and civilian leaders of the armed forces, the budgets of both the army and navy were to be reduced. Eager to realize a peace dividend, Republicans discounted the importance of feeding the vanquished Germans and Japanese, and of rebuilding their shattered economies. Not all GOP leaders supported such a self-defeating approach to the foreign aid budget; Senator Henry Cabot Lodge Jr., grandson of the man who had doomed Woodrow Wilson's League of Nations, compared his party colleagues to "a man wielding a meat ax in a dark room." But the vast hemorrhage of government money since prewar rearmament began had worried fiscally prudent Republicans, who remained focused on shrinking the size of the federal government. Persuading conservatives to provide money for Greece and Turkey in such an environment would not be easy.

One of Truman's aides later recalled that in his

dealings with the newly elected Republicans in Congress, "The president liked . . . face to face exchange. It personally meant a great deal to him. That kind of firsthand obtaining of the feelings, the attitudes, reactions, the dos and the don'ts from the congressional leadership coming straight to him, were, at least for him, by far the most effective way to do business."

So it was that on Thursday, February 27, in the Oval Office, Republican and Democratic leaders gathered in response to the president's summons. With Truman were Marshall and Acheson, with the former general taking the lead in laying out the crises now facing Greece and Turkey. The delegation included Arthur Vandenberg, the Republican chairman of the Senate Foreign Relations Committee; the Democratic ranking member, Senator Tom Connally; Speaker Joseph Martin; House minority leader Sam Rayburn; and the chairman of the House Foreign Affairs Committee, Charles A. Eaton. Top appropriators were also present. Missing was Senator Robert Taft of Ohio, son of former president and chief justice William Howard Taft, and a man who by ideology and temperament would be unlikely to support the president's proposal. In a glaring oversight, the White House had failed to invite him, a serious mistake given his influence in the party and especially clumsy considering his well-known isolationist views. Even without Taft present (and his absence was noted by Vandenberg), Acheson felt "we were met at Armageddon."

It made sense to give the secretary of state the starring role on that critical day. No less a figure than Winston Churchill had lauded General George Marshall, who had been chief of staff of the army in World War II, as "the organizer of victory." He was a lifelong servant of the state, drawn out of retirement by Truman to be the nation's chief diplomat. Austere and forbidding, he was held in awe by his subordinates, and his clout on the Hill was far greater than that of the president. The jocular Franklin Roosevelt had once ended an Oval Office meeting by turning to Marshall and asking, "Don't you agree, George?" Marshall had icily replied, "I am sorry, Mr. President, but I don't agree with that at all." Roosevelt was a man not used to being challenged, and never referred to the general by his first name again. But when the time came to appoint a chief of staff, FDR elevated Marshall over more senior colleagues, knowing that he would always get the truth from him.

For whatever reason, Marshall's quiet magnetism failed him on that momentous afternoon. After the president asked him to speak, the general read from his notes in a rote and uninspiring manner: "A crisis of the utmost importance and urgency has arisen in Greece and to some extent in Turkey. This crisis has a direct and intimate relation to the security of the United States." Marshall agreed with the recommendations of his staff, and stuck to the prepared script, but obviously was not yet inspired by Acheson's plan. His listeners grew restless; one asked, "Isn't this pulling British chestnuts out of the fire?" Old frustrations and antagonisms ran deep and resurfaced in that Oval Office meeting; more forceful response would be required to overcome congressional skepticism.

Sensing that they were losing the leadership of Congress, and trusting in the wisdom and humility of his chief, Acheson quietly asked Marshall if he might take over the presentation. "This was my crisis," he later remembered. "For a week I had nurtured it." Acheson then launched into a passionate soliloquy, shrewdly but sincerely playing on the anticommunism bent of his audience. "Soviet pressure on the straits, on Iran, and northern Greece," he warned, "had brought the Balkans to the point where a highly possible Soviet breakthrough might open three continents to Soviet penetration." Prefiguring later talk of nations falling like dominoes, he said, "Like apples in a barrel infected by one rotten one, the corruption of Greece would infect Iran and all to the east." And the infection would keep spreading. "Freedom itself" was at stake. There was no time to lose, for the Greek government was badly weakened.

A silence fell. Finally Senator Vandenberg said, "Mr. President, if you will say that to the Congress and the country, I will support you and I believe that most of its members will do the same." Such sentiments being openly expressed by an avowed isolationist like Vandenburg was an unexpected turn of events. Sensing a shift in momentum, other congressional leaders voiced their assent, at least in principle. So far, it seemed that "the nonpartisan oil of government lubricated the machinery of politics through the leadership," as Acheson later put it. But Truman and his team would soon learn how far they still were from resetting 150 years of US foreign policy.

Chapter 5
The Man from Missouri

I t was a hot August day in Kansas City, Missouri, but Harry Truman shivered as he stood naked before a bored army doctor. The usual indignities that come along with such an exam followed, but to Truman the discomfort was worth it.

The year was 1917, and the United States had officially entered the Great War in April. Truman wanted to serve and, for the second time in his life, joined the Missouri National Guard. But his unit was being called up for active service as part of the regular army, and regulations required that he pass a physical.

From the neck down, it wasn't a problem. The doctor noted that at five feet eight inches tall and weighing 151 pounds, his "figure and general appearance" were "average," and that his bones and joints, skin, and nervous and respiratory systems were "normal."

But his vision was anything but. When he was a boy, Truman was often taunted as a "sissy" in part because of his thick spectacles, without which the world was a shapeless blur. Though he may have tried to memorize the eye chart ahead of time, when he took off his glasses the letters dissolved into a series of indecipherable blobs. Whatever feat of memorization he may have attempted, the doctor was not fooled. His right eye was moderately weak, but his left eye had 20/400 vision. Next to this entry, the examiner scribbled, "Blind."

Somehow, and to his immense relief, he passed the examination. Perhaps the doctor had faith that Truman's spectacles would remain firmly in place in the trenches of France. Or maybe he saw the determination in those half-blind eyes.

Harry Truman was already thirty-three, much older than the average recruit. But though he was a man with a farm to run, a mother and sister for whom he felt responsible, and a sweetheart with whom he'd been in love since childhood, resisting the call of duty was not in him. Perhaps the call of adventure was powerful for this small-town farmer, but Truman remembered his heart being stirred by President Wilson's vow to make the world "safe for democracy."

Not even the most imaginative dramatist would cast Harry Truman in the role of world statesman. No president since Lincoln had arrived in the White House with so little formal education, and like that predecessor, he came from simple origins. Truman was not born in a log cabin, but his rural childhood in Missouri was not very different from Lincoln's in Kentucky. There was no running water or electric light, and silence at night was nearly unbroken. The reverberations of the Civil War over which Lincoln presided were still heard in the Missouri of Truman's birth, and both of his parents had inherited Confederate sympathies.

Harry S. Truman was born in 1884, within a decade of such twentieth-century giants as Franklin Roosevelt, Dwight Eisenhower, George Marshall, and Douglas MacArthur. But unlike them, he showed little early promise and lacked the propulsive quality and sense of destiny that drove those other political and military giants. In time he would take his place in the pantheon of architects of the American Century, but for him the path to greatness was long and dimly lit.

The nearsighted young Harry had a passion for books, and those thick spectacles that brought him such grief in the school yard also allowed him to indulge that love of reading. An early move to the nearby town of Independence helped expand his horizons and improve his education, but his thick glasses, poor health, and a lack of physical grace kept him on the outside of social circles in and out of school. He steeped himself in history, and in the stories of the past he learned about a world that stretched far away from rural Missouri. Later he would observe, "The only thing new in the world is the history you don't know."

His other youthful passion was classmate Elizabeth "Bess" Wallace, a striking girl whose comparatively well-to-do mother scorned Truman as a young man of no promise. He courted "Bess" with a dogged determination that eventually won her heart, but her mother would never improve her opinion. She remained convinced that Harry would never be a success, and even living in the White House left her unmoved in her opinion of Truman.

By the time he graduated from high school, his father had gone broke, so college was out of the question. Truman had hoped to attend West Point, but his poor vision made him ineligible. Instead he worked a series of odd jobs and helped his father manage the family farm. Harry was dutiful but bored, and the future before him held little promise. But he remained enchanted with the past and Andrew Jackson was a personal hero. Still, he disavowed any interest in a political career, writing that "Politics sure is the ruination of many a good man . . . to succeed politically [a man] must be an egotist or a fool or a ward boss tool."

Truman would instead respond to his country's call to arms. In the summer of 1914 Europe was ablaze. The assassination of the Austrian archduke Franz Ferdinand in Sarajevo on the twenty-eighth of June had sparked a continental catastrophe. The outraged Austrians, with the backing of their German allies, presented an ultimatum to Serbia that was more a pretext for attack than a realistic set of demands. On July 29 Austria-Hungary attacked Serbia, and the Russians mobilized for war to protect their fellow Slavs. Germany unleashed its long-planned invasion of Russia-allied France, marching first through neutral Belgium. Alarmed by the prospect of a German-dominated Europe, and obligated by treaty to preserve Belgian neutrality, a reluctant United Kingdom declared war on Germany on August 4. The Old World seemed set to devour itself, and from across the broad Atlantic, Americans watched with horrified fascination.

Under President Woodrow Wilson, the United States declared its neutrality and kept out of the fray. Statesmen in London might have dreamed that America would intervene on their side, but the United States' vast immigrant populations from Ireland and Germany had no interest in fighting alongside the British. If a decadent Europe chose to sacrifice its civilization on the altar of militarism, there seemed little point in the United States interfering. In 1914, the New World was thriving economically and growing exponentially. Becoming distracted by a distant archduke's death seemed irrational at best.

Germany predictably overreached. In May 1915 a German U-boat sank the *Lusitania*, a British ocean liner, off the coast of Ireland. Among the nearly 1,200 passengers killed were 128 Americans. Public opinion was inflamed, and anti-German feeling was widespread, but Wilson clung to his high-minded policy of nonintervention. America, he declared, was "too proud to fight."

Wilson may have been too proud to fight then, but the German outrages that followed ensured that America would soon have little choice but to strike back. In early 1917, the Germans sent a secret communication to the Mexican government known as the Zimmerman Telegram. In it, the Germans asked Mexico to join the fight against the United States, and promised that in return, Texas, Arizona, and New Mexico would be restored to them. In one of the great intelligence coups of the twentieth century, the British intercepted and decoded the message, and made it public.

Just before the contents of the telegram were revealed, the Germans resumed unrestricted submarine warfare and sank American merchant ships. These reckless provocations proved to be too much for Wilson and Congress to ignore, and the United States ended its splendid isolation with a declaration of war in April. In his war message to Congress, Wilson declared: "The world must be made safe for democracy," and in his ringing peroration uttered words that would find an echo in President Truman's three decades later:

It is a fearful thing to lead this great peaceful people into war, into the most terrible and disastrous of all wars, civilization itself seeming to be in the balance. But the right is more precious than peace, and we shall fight for the things which we have always carried nearest our hearts—for democracy, for the right of those who submit to authority to have a voice in their own governments, for the rights and liberties of small nations, for a universal dominion of right by such a concert of free peoples as shall bring peace and safety to all nations and make the world itself at last free.

Among the millions who would heed the president's call was the thirty-three-year-old Harry Truman, who with a series of business failures and bad investments behind him would don a soldier's uniform. The man who would bring the Second World War to an end with the atomic bombing of Hiroshima and Nagasaki would fight in the First as an obscure artillery officer. After six months of training at a bleak army camp in Oklahoma, Harry Truman embarked for Europe as a captain.

His service in France was honorable and there he displayed a yet-to-be-seen aptitude for leadership. Promoted to command of Battery D, he eventually forged a rowdy and rebellious gaggle of Irish Catholic troops into a potent fighting force. Their assigned section was different from most—rather than a flat and blasted landscape, they fought in the Vosges Mountains. But while the shape of the terrain was unique, the muddiness mirrored the rest of the front. Truman and his men exchanged murderous volleys with the Germans and experienced the terror of battle for the first time. From there they were dispatched to the Argonne Forest, from which they began an epic assault on the enemy lines. Several weeks later, German morale collapsed and the armistice was declared. The war was won and Truman had played a small but honorable part in that victory.

The pace of the newly minted war hero's life accelerated when he returned home. He and Bess were finally married, and he traded the life of a farmer for that of a businessman. Truman and his wartime sergeant Eddie Jacobson opened a haberdashery in Kansas City with high hopes for the future. But the luck that sustained him in the Argonne Forest deserted him in Kansas City, and a recession forced the partners out of business.

Thus it was that thirty-five-year-old Harry Truman found himself a vet who was both married and financially broken. But rather than proving to be the ruination he once predicted, politics would soon prove his salvation. An army friendship forged a connection with a local political boss who offered him the chance to run for Jackson County judge (commissioner). To the surprise of all—not least himself—Truman accepted, and one of the unlikeliest political ascents in American history began.

Harry won the judgeship and proved himself an adept administrator and retail politician. In an era marked by widespread graft and corruption, Truman won a reputation for honesty and good government that he would later display in the United States Senate. After a brief setback, he ascended to the post of presiding judge on the strength of that reputation (and with the help of the party bosses). By expanding and improving the county's road system, he helped lay the groundwork for future prosperity. Later, the Depression would predictably add great burdens to his efforts, but he handled the county's straitened circumstances efficiently and effectively. Truman was proving to be a strong leader and competent administrator.

In Missouri, political machines governed all, and the bosses decided which politicians would rise and which would fall. When a seat in the United States Senate became available in 1934, the Kansas City machine, headed by boss Tom Pendergast, decided that Harry Truman should be its candidate. Truman was taken aback by the offer, but could not resist the challenge. He hurled himself into the race, facing two opponents in the Democratic primary with years of national political experience between them. His efforts—and those of the Kansas City machine—paid off with a surprising victory. At the age of fifty, Harry Truman was on his way to the United States Senate.

Adjusting to life in Washington proved difficult for the man from Missouri. He never fully escaped the taint of the Pendergast machine, and some of his more fastidious new colleagues wanted little to do with him. But Truman, always a hard worker, devoted himself to Roosevelt's New Deal and gradually earned the respect of his fellow senators. The 1939 conviction and imprisonment of Boss Pendergast damaged his reputation and clouded his prospects, but Truman's essential decency (and good luck) allowed him to survive that scandal. Six years later, he would be president of the United States.

Chapter 6
Gnawing Away at Greece

American newspapers and magazines enjoyed a prestige and influence in 1947 that is difficult to imagine in an era when most print publications are struggling to simply survive. In an age before cable news and the Internet, publications like *Time*, *Life*, and the still venerable *New York Times* were not only sources of news and information, but also institutions with considerable political clout. Wooing the print press would be a critical part of any effort by the Truman administration to influence Congress. If the press lent their prestige to a policy, many on the Hill would have little choice but to follow their lead. Dismissing media outlets as "fake news" or villifying journalists as "enemies of the people" was not an option in a time when such autocratic ravings were reserved for tyrants like Joseph Stalin.

Although the drama of the past week had mostly been kept from the press, reporters were beginning to grow curious about events around the White House. The burst of activity in the State Department had been evident, and soon word would leak to the media. Truman believed it was time to satisfy their curiosity, and if possible, win editorial support.

Hours after the meeting with congressional leaders in the White House, General Marshall summoned a group of correspondents and shared some of the week's events. Everything was "on background," and reporters knew that to violate Marshall's rules by identifying the source would lead to their banishment from the inner sanctum. The secretary of state explained the strategic significance of Greece and Turkey and emphasized the urgency of the situation. More off-the-record briefings followed, and awareness of the crisis began to reach to the general public.

The overture to the press bore immediate fruit the following morning, February 28, in the form of a front-page story by the *New York Times*' legendary national correspondent James Reston. The Scotland-born Reston, age thirty-seven, was immensely influential and had won a Pulitzer Prize for his coverage of the

Dumbarton Oaks Conference that led to the creation of the United Nations. He had also played a prominent role in the American wartime propaganda effort by establishing the London branch of the Office of War Information, and understood the art of preparing the public for unexpected events that lay ahead.

Rather than peacetime propaganda, Reston's story in the *Times* was a faithful summation of the previous day's events; Reston also provided readers a powerful summation of the matter at hand. He observed, "There is far more at stake in this crisis than the appropriation of money or the economic plight of Great Britain. . . . In choosing to put economic stability first and in turning to the United States for assistance, the British have in effect asked whether the United States was prepared to assume a great part of the responsibility for world peace and stability assumed by Britain in the nineteenth century."

Reston breathlessly reported that the meeting with congressional leaders was "surrounded with unusual secrecy. Those who attended were sworn to silence and every effort was made to prevent the subject under discussion from being divulged to any more legislators than necessary to gain support for a loan."

But the word was now out, and the administration's efforts seemed to be proceeding like clockwork. On the morning of the *Times* story, Acheson assembled a larger group of State Department officials to inform them of the new policy's details and direct them how to best implement the administration's anti-Soviet plan. He told the group of the surprisingly supportive response the plan had received from congressional leadership the day before, and cautioned the group to avoid any direct criticism of the Soviet Union. Joseph Jones recalled that "it seemed to those present that a new chapter in world history had opened, and they were the most privileged of men, participants in a drama such as rarely occurs even in the long life of a great nation."

At that, Acheson departed the room and left Henderson and Hickerson to carry on the meeting. There were countless assignments to dole out, including the drafting of legislation and crafting of official responses to the British notes. Hickerson rallied the assembled diplomats by declaring that the situation in the eastern Mediterranean "was certainly the most important thing that had happened since Pearl Harbor," and then set the State Department officials to work.

Later that day, the director of the Office of Public Affairs, Francis Russell, convened a meeting with his counterparts in the Navy and War departments. The group's elaborate title was the "Subcommittee on Foreign Policy Information of the State-War-Navy

Coordinating Committee." Policy personnel from all three departments were also present. Their task was onerous: how to craft a message that would convince a skeptical Congress and public to support a major new foreign policy initiative that would have profound and long-lasting effects. While press outreach had already begun, a much more elaborate approach would be required to sell this dramatic policy shift to the nation.

It was agreed that little should be said about the British. The sour reference to "pulling British chestnuts out of the fire" at Marshall's congressional meeting was symbolic of wider public sentiment. And it would be unseemly to appear triumphalist as the once-powerful empire continued its rapid decline and contraction. Rather, the public relations campaign would present Truman's policy as one of supporting democracy throughout the world for the sake of American national security. There was a sense of exhilaration around the table at the sheer audacity of what the administration was attempting to achieve on the world stage. American isolationist sentiment had been so strong for so long that the government was accustomed to acting incrementally whenever foreign entanglements were involved in a new policy. Roosevelt had used every political skill imaginable to coax the American people into supporting the Allied cause long before Pearl Harbor, using elliptical phrases and homely metaphors to camouflage his true intent. But now a new and historic step was to be taken, and would soon be expressed in Truman's blunt and unambiguous style.

The subcommittee soon produced a report that would serve as a blueprint for the government's public relations policy in the weeks to come. Much of it would lay the foundation for the president's eventual speech to Congress. Drafted primarily by Russell, the report was focused, direct, and drafted with powerful prose.

INFORMATIONAL OBJECTIVES
AND MAIN THEMES
Basic United States Policy

A cardinal objective of United States foreign policy is a world in which nations shall be able to work out their own way of life free of coercion by other nations. To this end the United States has just finished fighting a war against Germany and Japan who were attempting to impose their will upon other nations. To the same end, the United States has taken a leading part in establishing the United Nations which is designed to make possible freedom and independence for all of its member nations.

The intent of this country to maintain a world of free peoples is directed equally against aggressive movements and against the imposition through whatever means from without of dictatorial regimes whether fascist, nazi, communist, or of any other form.

This principle of our foreign policy recognizes that only in such a world can the United States maintain its freedom and security.

A frank appraisal of the present world situation requires a recognition of the fact that a number of the countries of the world either have had forms of government imposed upon them against the will of a majority of the people or are in imminent danger of such a fate.

There is, at the present point in world history, a conflict between two ways of life. One way of life is based upon the will of the majority, free institutions, representative government, free elections, guarantees of individual liberty, freedom of speech and religion, and freedom from political oppression. The second way of life is based upon the imposition of the will of a minority upon the majority, upon control of the press and other means of information by the minority, upon terror and oppression. Such minority terroristic groups have various objectives. They may seek a fascist, a feudal, a communist or other order. But the major issue that is posed for the world is not one of objectives, not one between socialism or free enterprise, not one of progress or reaction, not one of left versus right. The issue is one of methods: between dictatorship and freedom; between servitude of the majority to a minority and freedom to seek progress.

The defeat of the axis powers was a milestone in the struggle for freedom. The end of the war, however, did not resolve the issue for all time. It is, in fact, alive in several areas of the world at the present time.

It is the policy of the United States to give support to free peoples who are attempting to resist subjugation from armed minorities or from outside forces. The United States will, within the framework of the Charter of the United Nations, assist in assuring the ability of peoples, who are now free, to work out their own destiny.

This is not a new policy. It was stated in the Atlantic Charter and in the Declaration of the United Nations, and was carried forward in the Yalta Agreement.

Through the granting of economic assistance and otherwise, we intend to help the Greek nation to preserve its free institutions.

This assistance should of itself give encouragement to other free nations through the notice that will thus be served that the United States recognizes the interdependence of all free countries.

A policy based upon the interdependence of free peoples does not necessarily betoken an increase in world tension nor an approach to war. On the contrary, the possibility of war will be greatly lessened. The continuing solidarity and strengthening of the free nations of the world will give support to the United Nations and thus strengthen the foundations of peace.

The free countries of the world, whether free enterprise or not, can co-exist peacefully provided there is no plan of conquest, domination or infiltration by any of them. The United States desires earnestly to effect with the Soviet Union a thoroughgoing understanding that will promote such a peaceful living together. It hopes and believes that this can be done.

The granting of economic assistance to Greece is consistent with the wholehearted support which the United States is giving to the United Nations.

Steps taken by the United Nations to promote reconstruction and insure the stability of nations has proceeded upon assumption that there would be inter-governmental economic assistance. The United States will continue to support and work through the United Nations in every way possible.

The present power relationships of the great states preclude the domination of the world by any one of them. Those power relationships cannot be substantially altered by the unilateral action of any one great state without profoundly disturbing the whole structure of the United Nations. Though the *status quo* is not sacred and unchangeable, we cannot overlook a unilateral gnawing away at the *status quo*. The Charter of the United Nations forbids aggression, and we cannot allow aggression to be accomplished by coercion or pressure or by subterfuges such as political infiltration.

The national security of the United States depends to a large degree on the maintenance of the principles of the United Nations and on maintaining the confidence of other nations in these principles. A seizure of power by a Communist minority in Greece would seriously impair that confidence.

On March 3, Paul Economou-Gouras, charge d'affaires of the Greek embassy, delivered the formal note from Greece requesting American assistance. The contents reflected suggestions provided by the State Department, and the tone of the message was urgent. It opened with grim references to "the systematic devastation of Greece, the decimation and debilitation of her people and the destruction of her economy," and after detailing the economic and military assistance it so desperately needed, declared, "The need is great. The determination of the Greek people to do all in their power to restore Greece as a self-respecting, self-supporting democracy is also great; but the destruction in Greece has been so complete as to rob the Greek people of the power to meet the situation by themselves."

But would the American people, and their representatives in Congress, after generations of instinctively resisting involvement in the affairs of other nations during peacetime, respond in time?

Chapter 7
Steer Clear of the Foreign World

Nearing the end of his second term, a weary George Washington issued a public letter entitled "The Address of Gen. Washington to the People of America on His Declining the Presidency of the United States." Despite a clamor for him to stand for a third term as president, Washington wished nothing more than to return to his beloved Mount Vernon. Having endured the slings and arrows of the highest office for eight years, he believed it was time to make way for others. By leaving voluntarily, he would establish a tradition of presidents serving no more than two terms that would endure for nearly a century and a half.

Drafted with the help of Alexander Hamilton, the letter was a profound reflection on Washington's time in office and the first president's appeal for national unity.

Known today as Washington's "Farewell Address," the message warned against the foreign intrigue that had threatened domestic harmony during his years in the presidency. Partisans of Britain or France, personified in Washington's cabinet by Hamilton and Thomas Jefferson, respectively, had allowed their preference for one country or the other to affect their political judgment and inflame their political followers. With consummate skill, Washington had steered the young republic through the diplomatic and military storms that raged across the European continent. In time, the United States might develop into a military power capable of contending with the empires of Europe, but Washington believed Americans should focus on their own domestic affairs, and take advantage of the protection provided by the expansive oceans that surrounded them. He wrote:

The great rule of conduct for us in regard to foreign nations is in extending our commercial relations, to have with them as little political connection as possible. So far as we have already formed engagements, let them be fulfilled with perfect good faith. Here let us stop.

Europe has a set of primary interests which to us have none; or a very remote relation. Hence she must be engaged in frequent controversies, the causes of which are essentially foreign to our concerns. Hence, therefore, it must be unwise in us to implicate ourselves by artificial ties in the ordinary vicissitudes of her politics, or the ordinary combinations and collisions of her friendships or enmities. . . .

Why, by interweaving our destiny with that of any part of Europe, entangle our peace and prosperity in the toils of European ambition, rivalship, interest, humor or caprice?

It is our true policy to steer clear of permanent alliances with any portion of the foreign world. . . .

The American people took this advice to heart, and while their isolation in the nineteenth century was never absolute, and the War of 1812 would bring them again into armed conflict with Great Britain, for the most part the United States remained aloof from the affairs of Europe. The population grew exponentially and the republic quickly expanded across the continent, in large part because that champion of limited government, Thomas Jefferson, nearly doubled the size of the country with the Louisiana Purchase. America's original sin of slavery deepened divisions between the North and South, resulting in the conflict the founders had assidously avoided. But while jaded European statesmen watched with interest as the upstart republic began tearing at the seams, and commercial ties between England and the Southern states might have led to foreign intervention in the "vicissitudes of her politics," the United States continued keeping foreign affairs at the margins of its political concerns.

Jefferson's warnings from his "Notes on the State of Virginia" that God's justice would not sleep forever on the issue of slavery were realized in 1861 when a civil war exploded over the issue. The Union's failure to put down the Confederate rebellion made European intervention in the American conflict a possibility. But deft diplomacy by President Abraham Lincoln and his secretary of state, William H. Seward, discouraged European powers from aggressively exploiting the American tragedy. The peace at Appomattox and the gradual reunification of the country set America on a course of mighty economic expansion, as the energy hitherto poured into sectional conflict was instead devoted to industrial development. Within two decades of the war's end, that expansion would usher in the Golden Age and a century of explosive economic growth.

Europe looked with wonder at the United States and the loud, clanging vibrancy of its economy and culture. A new aristocracy of American titans guided America's economy and influenced its politics in a manner that precipitated the development of a new unbridled capitalism. An era of relative peace on both sides of the Atlantic allowed America to further heal the wounds of war without foreign distractions. Only in 1893 did American ministers abroad first become identified as ambassadors, and while distinguished figures such as John Hay enjoyed enhanced status in London and elsewhere, diplomacy remained a leisurely craft that rarely drew attention from the population at large. Affluent Americans began traveling to Europe in far larger numbers, and interest in the Old World increased with Americans' disposable income. But a newfound interest in European culture and civilization did not translate into political support for a more engaged US foreign policy toward the Continent.

As Dean Acheson observed, "The impingement in the nineteenth century of what the Supreme Court has called 'the vast external realm' upon American interests occurred rarely, and usually only when wars between foreign nations interfered with our commerce or when foreign nations intervened in our hemisphere."

The first cracks in this isolationist resolve began to appear near the close of the century. The heavy-handed efforts of Spain to subdue its restive colony of Cuba inflamed the press and inspired ambitious war hawks in

Washington to prepare for battle. Many blamed the explosion of the warship USS *Maine* in Havana Harbor on the Spanish, and soon war fever gripped America. None celebrated the outbreak of the "Splendid Little War" more than the young assistant secretary of the navy, Theodore Roosevelt, who mobilized the US fleet on his own initiative.

President McKinley, whom Roosevelt derided for having "a backbone like a chocolate éclair," was inclined to caution. But the impatient assistant secretary would not be deterred, and the navy was deployed to battle the Spanish fleet. The resulting war was brutishly short, and gave the United States its first taste of possessing an overseas empire with holdings in the Philippines and Guam.

Still, the overwhelming preference of the American people was to look inward. Roosevelt entered the White House three years later, but was more restrained in the realm of foreign affairs as president than he had been under McKinley. Rather than starting wars, he gained international recognition for brokering a peace: After the Russians and the Japanese had come to blows in 1904, Roosevelt offered to mediate an end to the conflict. Thus did the "war hawk" become the first president of the United States to win a Nobel Peace Prize.

The first large-scale American military intervention abroad was its belated entry into the First World War in 1917, after repeated German provocations made neutrality politically impossible. But even that momentous engagement in the cauldron of Europe, which saw three million Americans in uniform and 114,000 killed, was quickly followed by a return to reflexive isolation. President Woodrow Wilson, the reluctant commander in chief, had been transformed into a passionate prophet of international cooperation, and traveled the country to promote American entry into the new League of Nations. Most Americans, and their elected representatives, remained unmoved by his exhortations, and the wartime president only succeeded in offending Republicans and wrecking his health. Wilson finished out his final term a near invalid.

Then as later, the Senate would prove to be an impregnable citadel of isolationism. In a foreshadowing of the events of 1947, a Republican chairman of the Senate Foreign Relations Committee would emerge as a key figure in Wilson's campaign for a more engaged foreign policy. But rather than the bipartisan champion to be discussed later, Henry Cabot Lodge Jr., of Massachusetts, opposed Wilson's draft treaty and allowed only a watered-down version to reach the floor. Inspired by near-religious zeal, Wilson refused to compromise, and instructed his supporters to oppose Lodge's bill. It was consequently rejected, and the president's preferred version was then voted down by an even greater margin. Well into the twentieth century, even after America's participation in a world war, Washington's admonition held sway.

For Americans, war remained a lamentable interlude, not a way of life. The military establishment that had so quickly expanded during its engagement in the First World War quickly contracted to its customarily modest proportions. The two decades that followed the 1919 Versailles peace agreement saw America's armed forces in dramatic decline. By 1940, the army was only the eighteenth largest in the world.

When Hitler launched his blitzkreig against Poland on September 1, 1939, Europe again was thrust into war; the United States under President Franklin Roosevelt again adopted a position of neutrality. So adamantly was public opinion against involvement in the latest European cataclysm that even the nightly reports of the merciless German bombing raids on London failed to ignite any martial spirit in the American electorate. The relentless courting of Roosevelt by the British prime minister, Winston Churchill (who succeeded Neville Chamberlain on May 10, 1940), and the ominous threat a triumphant Hitler might pose to the world, moved FDR to quietly favor the Allied war effort. But much to Churchill's frustration, Roosevelt moved with extreme caution and, as late as November 1940, while campaigning for an unprecedented third term, declared, "The first purpose of our foreign policy is to keep our country out of war."

His caution was justified. As late as September 1941, former president Herbert Hoover observed that after the First World War, "Europe degenerated into a hell, the brew from which poisons the earth today," and warned against "sacrificing our sons" in another war resulting from "the eternal malign forces of Europe." This apocalyptic language from Roosevelt's predecessor accurately reflected the feelings of millions of Americans about the Old World and its endless conflicts.

But for the Japanese's attack on Pearl Harbor and Hitler's foolish declaration of war on the United States four days later, American entry into the Second World War would have come much later, if it came at all. Only Churchill's deft personal diplomacy persuaded the Americans to focus first on the European theater rather than the Pacific. But the end of neutrality and the mobilization for total war were acts of self-interest on America's part, and not an altruistic campaign by US soldiers to save the British Empire from collapse.

There was every reason to believe that once Germany and Japan were defeated, the United States would retreat once again behind its geographical fortifications and forsake any further involvement in European affairs.

The historic tides created by the pull of World War II, and the scale of tragedy resulting from that great war, demanded a different response from Washington. The Soviet Union, wartime ally of the United States and the United Kingdom, had paid a disproportionate share of the butcher's bill for victory over Hitler. Nine out of every ten German soldiers who died in the war were killed on the eastern front at the hands of Soviet soldiers, and Russian casualties were horrendous. As many as twenty-nine million Russian citizens perished during the war, while the US and the UK each lost about four hundred thousand souls. Driven by an understandable sense of entitlement to the spoils of war, and an overwhelming ideological drive to spread communism across Europe, the Soviet Union embarked on a policy of expansion.

Having sunk so much blood and treasure into winning a war against fascism, would the United States now leave the field to the equally repellent creed of Soviet communism, and a leader who had killed tens of millions of his own citizens before the war? World War

II had also killed millions in Europe, ravaged huge sections of cities, and left masses of people starving amid the ruins; tuberculosis and other diseases were rampant. Such wretched conditions bred communist influence, with false promises of equality and economic justice. The question facing Truman following the war's end was whether the world's sole economic superpower would turn a blind eye yet again to an ever-expanding tyrannical regime.

Chapter 8
Passing the Torch

General Marshall departed Washington for the Conference of Foreign Ministers in Moscow on March 5. He would be out of the country for several weeks, with meetings spread across Paris, London, and West Berlin. Before boarding his airplane, Marshall addressed a phalanx of reporters and photographers, making his first public declaration of the crisis. Gazing out at the press with his ice-blue eyes, vapor rising from his mouth as he spoke in the near-freezing air, Trumans secretary of state said what he believed was at stake. The *New York Times* reported that Marshall called the issue of "primary importance," and "so far reaching and of such tremendous importance" that a formal declaration of policy "could properly come only from the president himself." The chilly morning was a fitting backdrop as he delivered the grim news: the economy of Greece had "deteriorated to the point of collapse," and "in the light of the world situation, this is a matter of primary importance to the United States."

The key players in the White House and the State Department had been stunned by the notice of British withdrawal from their former role as guarantor of international stability. In Asia, Africa, and the Middle East, the British had maintained a vast sphere of influence with a mixture of deft diplomacy, an assumption of superiority, and occasional brute force. The British Empire could be cruel and exploitative, but had also spread the blessings of improved health care, sanitation, transportation, education, and the rule of law to some of the unlikeliest corners of the earth. Having been born, educated, and trained in a world largely shaped by Great Britain, it was difficult for Truman and his men to imagine the changes that were sure to come. Assumptions shaping international affairs for two centuries were suddenly outdated and of no use in the crisis now erupting between the United States and the USSR.

As surprising as the news of February 21 had been, US policy makers were given some warning that Britain's reign was coming to an inglorious end. The Second World War produced vast upheavals from 1939 to 1945, but the declaration of peace in August 1945 brought little respite. The Soviet Union was on the march, Attlee's England was in decline, and American leaders looked upon the Atomic Age's dawn with a mix of both exhilaration and fear.

Britain's parlous economic state was hardly a secret; only American loans had made it possible for the UK to finance its war effort, and it was obvious that their financial reserves were exhausted. The British foreign secretary, Ernest Bevin, and the then secretary of state James F. Byrnes had conferred on numerous occasions about Britain's economic and strategic challenges in the wake of the war.

Soviet ambitions had been set in motion. Like a shark smelling blood in the ocean, Stalin was ready to move on Britain's former colonies and clients. Truman and the British prime minister, Clement Attlee, knew the "Man of Steel" was not to be trusted. He had already reneged on his promise of free elections in Poland and as other Eastern European countries fell under the Soviet's shadow, this dictator would not hesitate to devour any territory that showed itself vulnerable. While the British and American people longed for a return to normalcy and a retreat from world affairs, Soviet leaders sensed a historic opportunity to launch outward offensives in all directions. The Second World War had exacted a sickening cost on the Russian people, and Stalin was determined that the USSR would never be invaded again.

The signs of Soviet aggression were most evident in Iran. The British and the Russians had occupied that oil-rich state in 1941 to keep the Germans from exploiting its resources. Under the terms of an agreement with the Iranians, the occupiers were to withdraw within six months of an Allied victory, and had committed to "respect the territorial integrity, sovereignty, and political independence of Iran." True to their word, the British had departed by the designated date, March 2, 1946. The Russians, meanwhile, remained in place. Having seized his prize, Stalin was not about to allow something as trivial as a treaty to dislodge Russian troops. Rather than withdraw, Stalin dispatched tanks and additional soldiers into the region. The British and the Americans looked on with alarm as these new formations moved west toward the borders of Turkey and Iraq. Brushing off diplomatic protests, the USSR relentlessly strove to extend its reach.

Thus, it was alarming but unsurprising when Stalin trained his eye on Greece and Turkey. The Soviet Union had long coveted unfettered access to the Mediterranean through the Black Sea, an ambition that Turkey had repeatedly thwarted. In August Stalin declared a desire to "defend" the Turkish straits, which naturally would have required a strong Soviet military presence in Turkey. Acheson sent a cable to then Secretary of State Byrnes recommending a stance that would find its ultimate expression in the Truman Doctrine:

In our opinion the primary objective of the Soviet Union is to obtain control over Turkey. We believe that if the Soviet Union succeeds in introducing into Turkey armed forces with the ostensible purpose of enforcing the joint control of the Straits, the Soviet Union will use these forces in order to obtain control over Turkey. . . . In our opinion, therefore, the time has come when we must decide that we shall resist with all means at our disposal any Soviet aggression and in particular, because the case of Turkey would be so clear, any Soviet aggression against Turkey. In carrying this policy our words and acts will only carry conviction to the Soviet Union if they are formulated against the background of an inner conviction and determination on our part that we cannot permit Turkey to become the object of Soviet aggression.

In a preview of future events, Truman conferred on August 15 with Acheson; Forrestal; and General

Dwight D. Eisenhower, then chief of staff of the army, about how best to deal with the Turkish crisis. Their options were made more complicated by the fact the Soviets had shot down unarmed US military transports during a dispute about postwar Italian and Yugoslavian territory. Speaking on behalf of his colleagues, Acheson recommended dispatching an aircraft carrier to Turkey as a response to the Turkish request for assistance and a demonstration of resolve.

When Truman agreed, Eisenhower raised a word of caution quietly asking whether those in the room were aware that such a decision could lead to war. In response, the president pointed to a map of the eastern Mediterranean and explained the importance of shielding it from Soviet infiltration. As Acheson recalled, "None of us doubted he understood fully all the implications of our recommendations."

The intervention worked, and the Soviets backed down—for a time. An insatiable Russia would continue to probe for weak spots throughout Europe and the world. But Soviet actions had left little doubt that the eastern Mediterranean would be one of their principal targets.

The issue presented itself again later in the year, after the Democrats' disastrous midterm election results. The Greek prime minister, Constantine Tsaldaris, visited Truman during a three-week trip to the United States in December, during which he was made an honorary citizen of New York. While in Washington Tsaldaris stayed in Blair House, the government's official guest house for foreign leaders, and conferred with the president, Byrnes, and Acheson. The prime minister recalled later that Britain had encouraged him to ask the Americans for help. Unfortunately, the Greek leader was so extravagant in his request that Acheson later dismissed him as "a weak, pleasant, but silly man"; Greece's opportunity to lay the groundwork for a serious first proposal was wasted. Truman felt more generous, having previously written to Acheson, "The Greeks were almost annihilated fighting our common enemy, the Germans, and while they have had some severe internal difficulty with the British, I can't help but feel extremely friendly to the Greeks." Warm feelings aside, Tsaldaris left Washington empty-handed. It would take a more decisive move from the British— and a more realistic set of policy proposals—to move the American machinery of government into action.

Truman needed more information before making a move on Greek aid. For this he turned to Paul A. Porter, the former chairman of the Federal Communications Commission. Porter was a lawyer and newspaperman from Kentucky who had been drawn to Washington to work in the Roosevelt administration. He began his government service as a legal assistant in the Department of Agriculture, and later served in senior positions in the Office of Price Administration and the Office of Economic Stabilization, both New Deal agencies. Roosevelt appointed him FCC chairman in 1944. In December 1946, just after his meeting with Tsaldaris, Truman tapped Porter to be the head of a new American Economic Mission to Greece, with the rank of ambassador. The *New York Times* reported, "It was regarded as a significant move by the United States in support of Greece in the face of the situation inherent in the fighting and chaos of the north." In a portent of things to come, the article explained that Porter would "consider, specifically, the extent to which the Greek Government can carry out reconstruction and development through effective use of Greek resources and the extent to which foreign assistance may be required."

Acheson announced the mission to the press and hailed the "valiant stand of the Greeks against the Nazi invasion, their continued resistance and sacrifice through the long occupation, and the hardships consequent upon the war which they have endured since liberation," and noted the "close relationship with Greece, particularly because of cultural ties between the two countries, and because of the large numbers of American citizens of Greek descent."

Porter was immediately struck by the grave conditions he discovered in the Greek countryside. In an article for *Collier's Magazine* months later, he related the story of a Greek peasant he encountered who shrugged and said, "Four times in my lifetime my home has been destroyed—by the Turks, the Bulgars, the Nazis, and the guerrillas. Why should I build it up again?"

After a month of on-the-ground fact-finding, Porter amassed information helpful to Truman based not only upon discussions with the Greek government—which he despised—but also with hundreds of Greek citizens across the ravaged country. In a report to the State Department, the new ambassador diagnosed "the unhealthy psychological condition of the people . . . a sense of helplessness on their part; a feeling that because they suffered during the War they should now be cared for by their richer allies; a belief that the external factors in their problem are so large that their individual efforts are futile."

He further lamented, "There is really no State here in the Western concept. Rather we have a loose hierarchy of individualistic politicians, some worse than others, who are so preoccupied with their own struggle for power that they have no time, even assuming capacity, to develop economic policy. The civil service is a depressing farce."

Porter had little to say about the communist threat. A committed New Dealer, he suggested in his reports that economic aid alone would be enough to maintain the Greek government's hold on power. Ambassador MacVeagh found those views narrow and naive, and urged him to take the problem of subversion more seriously. Porter reluctantly acceded.

Even before Stalin began showing his hand in Iran and Turkey, President Truman had received a powerful warning about Soviet designs in the unlikely setting of a college gymnasium in his home state of Missouri.

After five long years in power, the British wartime coalition came to an abrupt end with Labour's withdrawl in the spring of 1945. Winston Churchill was no longer the leader of a national government but now simply a Tory prime minister in a country eager for change. A general election was set for July—the first in ten years—with Berlin smoldering in ruins and the war in the Pacific raging toward an epochal end. Before the results of Britain's election was declared, Churchill joined Joseph Stalin and the new American president, Harry Truman, in Potsdam, Germany, for a conference to decide the immediate fate of the post-war world. Churchill's former wartime deputy, Labour leader Clement Attlee, also attended, in case the ongoing count should elevate him in Churchill's place. Few people, least of all Churchill, thought the defeat of the man considered the savior of the British Realm was possible, but it was with some trepidation that he returned to London in midconference for the declaration of the vote. Most in the prime minister's entourage, confident that they would be returning with a triumphantly re-elected prime minister, left their luggage behind.

It was not to be. The British people, exhausted by war, determined to punish the party of prewar appeasement, and eager to receive more expansive government benefits, voted for Labour in overwhelming numbers. The prime minister who had heroically guided his country to victory in the Battle of Britain was now unceremoniously booted out of 10 Downing. As the election results became clear, his wife, Clementine, remarked that the defeat might be "a blessing in disguise." He glumly replied, "At the moment it seems quite effectively disguised." A sympathetic King George VI tried to cushion the blow by offering Churchill membership in the Order of the Garter, the highest order of knighthood, but Churchill respectfully declined. With his customary humor he asked the sovereign, "How can I accept the Order of the Garter, when the people of England have just given me the Order of the Boot?"

Churchill was understandably shattered by the rejection, but with characteristic resilience and resolve he threw himself into writing memoirs, making good on his promise that history would be kind to him—for he would write it—and in the process, the former prime minister finally freed himself from a lifetime of debt. Churchill was far less keen on the mundane details of his new post as Leader of the Opposition, but the position was a necessary burden for the achievement of his fondest wish: a triumphant return to Downing Street.

Though out of power, Churchill still yearned to shape the course of world events. And in the escalating tensions between the Soviet Union and the West, a conflict he had foreseen while Stalin was still a wartime ally, he found a new cause and focus. The wartime alliance with the United States must be revived, Churchill believed, this time to thwart the threat of Soviet tyranny. A new conflict loomed—a cold war with a communist bloc that loathed Western democracy and sought to crush the free will of its subjects. Such an evil must be resisted, for after the sacrifices required to destroy Nazism, now was not the time to appease another tyrant.

But British foreign policy was fashioned by the government, led by the new prime minister, Clement Attlee. Though Attlee was a good and honorable man, Churchill believed him ill equipped to lead this crusade. The former prime minister would derisively dismiss his successor as "a sheep in sheep's clothing." With another general election far in the future, Churchill feared he had no time to lose. How could he regain his voice, and reinvigorate Britain's old alliance with the Americans?

The answer came in the unlikely form of a letter from the president of Westminster College in Fulton, Missouri, an obscure institution with a few hundred students, inviting him to accept an honorary degree and deliver a speech. Churchill would have declined the invitation under normal circumstances; then as today, Fulton is remote and inaccessible to many. But handwritten at the bottom was a postscript that must have made the old war leader's heart skip a beat: "This is a wonderful school in my home state. Hope you can do it. If you come, I will introduce you." It was signed "Harry S. Truman."

Laid before Churchill now was the extraordinary opportunity to deliver a speech in the American heartland that—however nondescript the venue—would command attention based on Churchill's own celebrity. More importantly, it was also a chance to take a long train journey from Washington to Missouri and back with the president of the United States. The former prime minister would have many hours to cement a relationship with the leader he had only briefly met at Potsdam, and to gain Truman's support for his postwar vision. Having the president by his side for the address would also imply American support for his stirring message. He swiftly accepted the invitation and the date was set: March 5, 1946.

In a lifetime filled with history-bending orations, Winston Churchill's speech in the Westminster College gym would loom large among them. Before a swarm of journalists and news cameras, Churchill uttered perhaps the first public declaration of the new Cold War: "From Stettin in the Baltic to Trieste in the Adriatic an iron curtain has descended across the Continent. Behind that line lie all the capitals of the ancient states of Central and Eastern Europe. Warsaw, Berlin, Prague, Vienna, Budapest, Belgrade, Bucharest and Sofia; all these famous cities and the populations around them lie in what I must call the Soviet sphere, and all are subject, in one form or another, not only to Soviet influence but to a very high and, in some cases, increasing measure of control from Moscow."

Truman publicly remained noncommittal publicly about Churchill's stunning Westminster speech—he denied, however implausibly, any knowledge of what his British guest was going to say. Though he was president of the United States, inside that Missouri gym Truman was a supporting player to the most famous statesman on earth. Did he feel, as the British journalist Alistair Cooke suspected, "wistful envy . . . in the immense shadow of Churchill?" Cooke thought later that the Truman Doctrine was conceived as the president listened to Churchill's call to action. One year later, it would be the American president offering his vision for confronting the new scourge of Soviet tyranny. Churchill would indeed return to Downing Street as prime minister in 1951, and focus his remaining energies on brokering peace with the Russians, but he and his country would play only a supporting role in the Cold War drama that was to come. Though none in attendance could have been aware of it at Westminster, the torch of world leadership had been passed from Churchill to Truman inside that college gym built on the dusty plains of central Missouri.

Chapter 9
A Baptism of Fire

Harry Truman had long been a skeptic of US isolationism. Years before he would address a joint session of Congress on the urgent need of America containing the growing Soviet threat, Senator Truman spoke to an audience in Larchmont, New York, on April 20, 1937, about "national defense and its relationship to peace." The former judge whose concerns had been limited to local matters such as building roads a few years earlier was now contemplating the approach of another world war. With admirable clarity, Truman warned, "We must not close our eyes to the possibility of another war, because conditions in Europe have developed to a point likely to cause an explosion at any time." It would be another four years before President

Franklin Roosevelt would speak with such clarity about US interests in yet another World War.

"We all want peace, and we all want to stay out of war, but we must go about it intelligently," he declared. He warned that the Neutrality Acts that Congress had recently passed were no guarantee of peace, but merely a "hope." He scolded the United States for having gone "hysterical" over disarmament in the wake of the First World War, and called for "an adequate Navy" and "an air force second to none." He invoked his hero Andrew Jackson, "the fighting old president from Tennessee," who said, "We shall more certainly preserve peace when it is understood that we are prepared for war." But Truman's warnings more resembled those that Winston Churchill had been sending the British people throughout the 1930s while languishing in the political wilderness where he would remain until 1940.

Reelected against the odds on the other side of the Atlantic that same year, Harry Truman later earned national prominence by organizing a far-reaching review of the government's military mobilization. America was technically at peace, but Hitler's blitz across Europe made it necessary for the US government to quietly begin preparing for a possible future conflict. Unprecedented sums of appropriated dollars were being transferred from the government into the pockets of defense contractors and third parties, and much was being wasted and mismanaged. The Special Committee to Investigate the National Defense Program, established more than six months before the Japanese attack on Pearl Harbor, was a bipartisan effort to ensure that government funds were not squandered. It was also an important opportunity for Truman, who became its chairman. Soon known as the Truman Committee, the panel launched investigations into every aspect of the defense industry, uncovering extensive waste and fraud. At Truman's first hearing, the committee heard from such witnesses as army chief of staff General George C. Marshall, whom the senator would come to revere and who would obviously play an important role in Truman's future. The blaze of publicity following those hearings would elevate the once-obscure senator into the public eye. In March 1943, Truman would appear on the cover of *Time* magazine, which was then considered a singular honor.

Much of what the junior Missouri senator uncovered was unflattering to the Roosevelt administration. Despite his passionate support for the New Deal and unfailing loyalty to the president, Roosevelt was often unhappy with the crusading senator. After one episode, Truman wrote of FDR, "He's so damn afraid that he won't have all the power and glory that he won't let his friends help as it should be done." But despite his frustration with the most patrician of presidents, Truman's value to both the war effort and the Democratic Party continued to rise.

The punishing schedule of the Truman Committee, with countless hearings on every conceivable subject involving defense, finally proved too much for its chairman. He was hospitalized due to exhaustion and ordered to cut back on his activities. But the crush of his schedule, as well as the coming war, would allow no rest, and Harry Truman threw himself again into the work of the committee. He was driven by a desire to serve his country, an absolutist's sense of right and wrong, and a near-maniacal desire to expunge the shame of his earlier connections to the Pendergast machine.

By the time of the 1944 Democratic National Convention in Chicago, which would nominate Franklin Roosevelt for the fourth time, Truman had become a player in the party but remained an outsider to the cadre of East Coast elite who surrounded Roosevelt. Two possible candidates for vice president, both of whom believed they were Roosevelt's first choice, asked the Missouri senator to nominate them. Truman committed to support former senator James F. Byrnes, but as FDR was leaving the final decision to the convention, any outcome seemed possible. The only result that seemed clear was that the left-wing incumbent, Henry Wallace, would be thrown off the ticket.

As the fateful decision drew nearer, events began to unfold in a way that Truman could not have ever imagined. Days before driving to Chicago, at a martini-fueled meeting with Roosevelt, a group of Democratic influencers considered the question of the number-two slot on the ticket. After hours of contentious debate, they finally settled on Truman. And so it was that Truman answered a knock on his hotel door in Chicago to find the chairman of the Democratic National Committee, Robert Hannegan, standing before him. To Truman's surprise and dismay, Hannegan told him that he was Roosevelt's choice. "Tell him to go to hell" was Truman's undiplomatic reply. But it was too late for protests from a reluctant politician; the wheels had already been set in motion and Truman was left with no choice but to accept FDR's call. That resistance crumbled completely when Roosevelt, a man who hated to be crossed, snapped loudly enough on the telephone to Hannegan for Truman to hear it from across the room: "Bob, have you got that fellow lined up yet?" Hannegan said no, calling Truman "the contrariest goddamn mule from Missouri I ever dealt with." With this, Roosevelt growled, "Well, you tell the senator that if he wants to break up the Democratic Party in the middle of the war, that's his responsibility." Truman took the phone and weakly protested further, but within minutes he bowed to the inevitable: "Yes, sir, I know you're commander in chief. Yes, sir. Yes, sir. Well, if that's what you want, that's what I'll do."

A last-minute boomlet for Wallace on the convention's second day, fueled by fake tickets and other tried-and-true dirty tricks, could not stop Truman's nomination. He was selected on the second ballot amid scenes of chaos and excitement. Robbed of suspense regarding the presidential nomination, Democrats across the convention hall entertained themselves by indulging in an exceptionally dramatic display for the VP race. Exhausted by a night of raucous campaigning, Truman had little to say by the time he addressed the Democratic delegates at Chicago Stadium, delivering what David McCullough called the shortest acceptance speech in American political history. Truman looked like a small and timid man in the glare of the spotlight, and the press soon dubbed him "the Missouri Compromise." For many like Indiana senator Samuel Jackson, Truman's greatet achievement at the convention was stopping the candidacy of the incumbent vice president. Jackson later said that he wanted inscribed on his tombstone the words, "Here lies the man who stopped Henry Wallace from becoming the President of the United States."

Upon returning to Washington on FDR's ticket, Truman resigned from the committee that had come to bear his name. He gathered his loyal aides and offered his sincere thanks for their dedication and hard work, and told them that they could be proud of having saved the government as much as $15 billion. The Truman Commission's success was a notable exception to Woodrow Wilson's observation that investigations of the executive branch by Congress "do not afford it more than a glimpse of the inside of a small province of federal administration. . . . It can violently disturb, but it often cannot fathom, the waters of the sea in which the bigger fish of the civil service swim and feed. Its dragnet stirs without cleansing the bottom." Wilson's protests aside, Truman's work proved to be an important step toward ensuring a more efficient war effort following Pearl Harbor.

The 1944 presidential campaign was a stormy one, with the war still raging and the president beaten down by illness and overwork. Roosevelt and Truman were a study in opposites, but both had a gift for politics and their own ways of attracting support. They encountered each other only rarely during the campaign, and what Truman saw filled him with dread. So frail was Roosevelt that he feared that the upcoming election would determine not only the next president, but the next two presidents. When a friend told him outside the White House that he would soon be its occupant, Truman replied, "I'm afraid you're right . . . and it scares the hell out of me." But Truman worked hard on behalf of the ticket, traveling 7,500 miles across the country delivering speeches on the back of his train. (Roosevelt had told him to avoid airplanes, because "one of us has to stay alive.")

The effort paid off. Despite predictions of a closer-than-usual race, Roosevelt and Truman handily defeated the Republican ticket headed by Governor Thomas Dewey of New York by 432 electoral votes to 99, winning the popular vote by more than 3.5 million. The American people had heeded yet again Lincoln's admonition—while he ran for reelection in 1864 during the Civil War—that it was "best not to swap horses in midstream." Roosevelt would be given the chance to preside over the coming victory, if only he could live long enough to see the war through to its conclusion.

The inauguration on January 20, 1945, was one of the smallest such ceremonies in modern times. With the war still being fought in the Atlantic and Pacific, and with the president ailing, there would be no traditional procession to the Capitol. Truman and Roosevelt took their oaths of office on the South Portico of the White House, speaking briefly in the bitter cold of that January day. Truman remained somber throughout the cereomony, contemplating the fate that appeared to be confronting him. The new vice president could only hope that Roosevelt would somehow find the strength to complete his fourth term. Unbeknownst to him, the president suffered searing chest pains just after the swearing-in, and had to fortify himself with whiskey just to greet guests. Had Truman been aware of just how frail FDR was, his gloom would have certainly deepened.

The new vice president's old benefactor Boss Pendergast died in St. Louis a few days later. For any number of reasons—the dignity of his office, concern for his political future, or justifiable caution—Truman might have remained in Washington and discreetly sent his condolences. But the Missouri man's overwhelming sense of loyalty and gratitude compelled him to attend Pendergast's funeral in person. Though politically unwise, Truman's decision foreshadowed the kind of fearless tenacity that he would later bring to the presidency. Harry Truman had a strong and innate sense of right and wrong, and decisions that would have caused anguish to other politicians rarely troubled him. His clear-eyed view of the world would guide him through some of the most momentous and terrible decisions ever made by a president.

The vice presidency that Truman had so disdained, and that he had done so much to avoid, lasted less than three months. As feared, Franklin Roosevelt had been reelected president of the United States a dying man. The first vice president to succeed to the highest office upon the death of the president, John Tyler, had been disdained as "His Accidency." But few in such a position could be considered as "accidental" as Harry Truman. Less than eleven years before he sat behind the Oval Office desk, Harry Truman had been the presiding judge of Jackson County, concerned with road building and his county's payroll. Now he would be in charge of bringing the greatest war in history to a successful conclusion, and building a lasting peace out of the ruins of Europe and Japan.

These awesome responsibilities fell upon him all at once, but none was as daunting as having to decide whether to use an atomic weapon against the Japanese people. In the barren hills of Los Alamos, scientists from around the world had harnessed the most elemental force in the universe, and with remarkable speed created a weapon that could level entire cities.

In an act of wanton recklessness, Roosevelt had kept the development of the atomic bomb a secret from his vice president and likely successor. Only after his first cabinet meeting as president, when the secretary of state, Edward Stettinius, asked him for a private word, was Truman informed of, as he later put it, "the development of a new explosive of almost unbelievable destructive power."

One of Roosevelt's priorities at the Yalta Conference in February 1945 had been to secure Soviet participation in the endgame against Japan. Though by then the US was only months away from its first and successful test of the atomic bomb, the president could not be certain that the superweapon would work. Conscious of Japan's fearsome capacity for resistance and the staggering casualties allied forces would likely endure in an invasion of Japan, FDR wanted the Soviets to play a part in the final stage of the Pacific war. And indeed they did, invading Japanese-occupied Manchuria in what has become known as Operation August Storm.

President Truman had been informed of the military's final plans for America's invasion of Japan in June 1945. A two-phase assault, beginning with a landing on the southern island of Kyushu, would require some six million men and was predicted to cost the lives of half a million Americans. As that invasion proceeded,

US bombers would continue their relentless assault on Tokyo and other cities, hoping to break their leaders' will with conventional weapons. In early March, a massive raid over the capital had dropped 1,665 tons of bombs, most of them incendiaries, and killed at least one hundred thousand people—more than died at Hiroshima or Nagasaki.

But even this nightmare from the sky did not move the Japanese toward a surrender. It seemed clear to US military planners that the enemy would fight to the bitter end, and that massive American casualties would accumulate before Japan's government was brought to its knees.

The first detonation of an atom bomb took place in New Mexico on July 16, a successful test that confirmed for Truman that he was "now in possession of a weapon that would not only revolutionize war but could alter the course of history and civilization." He was informed of the test just after his arrival in Potsdam, Germany, for the first meeting of the American, British, and Soviet leaders ("The Big Three") since Yalta. But even the news of a successful atomic test did not immediately change US plans for defeating Japan; the mighty invasion was still—officially, at least—"on the books." And the war was projected to last until the end of 1946. In an effort to convince the Japanese of the futility of further resistance, the United States, the United Kingdom, and China issued a stark warning: "We call upon the government of Japan to proclaim now the unconditional surrender of all Japanese armed forces, and to provide proper and adequate assurances of their good faith in such action. The alternative for Japan is prompt and utter destruction."

Truman later reflected, "The final decision of where and when to use the atomic bomb was up to me. Let there be no mistake about it, I regarded the bomb as a military weapon, and never had any doubt that it should be used." With the hope that the Pacific war might be swiftly ended, and that countless American lives might be spared, Truman issued the order to begin the atomic bombing of Japan after August 3, 1945.

On the sixth of August, 1945, a silvery Boeing Superfortress B-29 bomber called the *Enola Gay*, with an atomic bomb code-named "Little Boy" rose from a runway in the Northern Mariana Islands and arced its way northwest to Japan. The day was hot and clear, and after several hours of flying, pilots Paul Tibbets and Robert A. Lewis could see their target: the industrial city of Hiroshima, with a population of 345,000. At an altitude of more than 31,000 feet, "Little Boy" slid out of the bomb bay and fell toward the earth, detonating nearly 2,000 feet above the city. The results were as horrific as the bomb's creators could have imagined, and far more horrifying than any of those on the ground could have imagined. The equivalent destructive force of 16,000 tons of TNT tore through the city, leveling most of the buildings and killing 80,000 people, most of them civilians. Many more would die later of radiation sickness. Peering down from the *Enola Gay*, an awed Lewis scrawled in his logbook, "My God, what have we done."

Informed of the bombing of Hiroshima while sailing home from Potsdam on the USS *Augusta*, Truman said to those around him, "This is the greatest thing in history. It's time for us to get home." In a statement released soon afterward, the president warned that if the Japanese failed to surrender, "they may expect a rain of ruin from the air, the like of which has never been seen on this earth. Behind this air attack will follow sea and land forces in such numbers and power as they have not yet seen and with the fighting skill of which they are already well aware."

But Japan remained unmoved.

Three days later, another bomb, this one codenamed "Fat Man," was dropped by another B-29 over the city of Nagasaki, causing less damage than "Little Boy" but still instantly incinerating tens of thousands of people.

Japan finally recognized the futility in continued fighting, and surrendered on August 14. In a radio broadcast the following day, Emperor Hirohito uttered one of the great understatements of the twentieth century when he observed that the events had gone "not necessarily to Japan's advantage." The war was over, and when General Douglas MacArthur formally accepted the Japanese surrender on the deck of the USS *Missouri* in Tokyo Bay, he declared, "Let us pray that peace be now restored to the world and that God will preserve it always. These proceedings are closed."

No newly sworn-in president could have expected to receive such a baptism by fire as Harry Truman. But Truman's goal of "seeking a peaceful world," even through the most horrific means, had just begun with the defeat of Japan. A world left in ruins now had to be rebuilt, and as the only nation on earth to emerge from the bloody war both stronger and economically unscathed, it fell upon the United States to guide postwar reconstruction, and preserve the gains that had been won against Hitler's Axis powers. Leading the nation through that perilous journey was a man who had failed to keep a haberdashery in business, and whose education had ended after a year at Spalding's Commercial College in Kansas City.

It is tempting to idealize the humble beginnings of great leaders, and to see in their rise proof of America's exceptional promise. But while a person blessed with good fortune and a capacity for hard work can often change their circumstances, they cannot always leave behind some of the limits of their origins. Lincoln came close: though his rustic prairie manners alarmed and offended the genteel Washington elite, his refinement of spirit and magnanimity rendered him immune to petty squabbles.

But in the Oval Office, Truman proved quickly to those around him that he was no Lincoln. Even as president, Truman never lost the rough and combative streak born out of his hardscrabble background. Those mocking taunts of the "sissy" in thick spectacles developed in Truman a pugnacious personality later in life. His rhetoric could be crude and slashing, even when directed against a nominal ally. He was even more vituperative about those working to block his agenda. And when his daughter, Margaret, an aspiring singer, received an unkind review from Paul Hume of the *Washington Post* ("she is flat a good deal of the time. . . . Miss Truman has not improved in the years we have heard her"), the president infamously dashed off a furious letter to the critic, calling him "a frustrated old man who wishes he could have been successful," and continuing: "Some day I hope to meet you. When that happens you'll need a new nose, a lot of beefsteak for black eyes, and perhaps a supporter below!" After his 1948 election to the presidency in his own right, Truman fired the director of the Secret Service for having attended his opponent's abortive "victory party." This rhetorical excess and occasional coarseness of spirit would make the effort to secure support for the Truman Doctrine even more challenging. Its eventual success was a testament to the seriousness of the issues at hand and the ability of Congress to set partisanship aside—even temporarily.

In spite of these rough edges, however, the United States was blessed to have Harry Truman at the helm of its government in 1947. Though he was often unfavorably compared to men like Acheson and the other expensively educated diplomats who served under him, Truman's voracious appetite for reading and his profound historical knowledge gave the new president a perspective that others around him lacked. As New York's Senator Daniel Patrick Moynihan once told a young lawyer named Tim Russert after being asked what he could offer compared to the Ivy League grads who filled the senators office, "Son, you know things already that they will never know." Like the future *Meet the Press* legend, Harry Truman's mind was uncluttered by the constant equivocation and paralyzing doubt that often make intellectuals better suited for the academy than high office. As one of his aides observed, "There is . . . such a thing as being too intellectual in your approach to a problem. The man who insists on seeing all sides of it often can't make up his mind." And having ascended to the White House fully aware of his own limitations, Harry Truman stubbornly resolved to do his duty to the best of his ability and surround himself with the finest minds that Washington and the world could provide. As the events of 1947 came hurling toward Truman and his team, those assets would prove to be enough to make Harry Truman the greatest foreign policy president of the postwar era.

Chapter 10
A New World Order

Until recently, a presidential address was the work of a sprawling bureaucratic committee. Often, as in the case of the State of the Union, the staffs of government departments work in concert with the president's speechwriters to craft the speech, with great energies expended over the drafting of a single sentence or phrase. This process is usually contentious; clashing agendas between agencies and government actors across the city are exposed. Presidential time and attention have long been considered precious commodities, and the exposure resulting from such a speech has been known to launch a program or sustain an initiative that would otherwise die from lack of public support. The battle among federal agencies for the

ear of the president has long been a continuous reality around Washington's bureaucracies. Actors within
the realm of foreign affairs, the Departments of State
and Defense—along with the National Security Council and intelligence agencies—often fight the hardest to
place their policy priorities in a president's address.

But the drafting of the speech to launch the Truman
Doctrine was the Department of State's responsibility
and its leaders' moment to shine. Under its revered and
gifted deputy, State would finally be afforded the opportunity to move past its reputation for pinstriped pre-
varication. Joseph M. Jones, a State Department official
who helped shape the events of 1947 and record them
for posterity, reflected later: "All . . . were aware that
a major turning point in American history was taking
place. The convergence of massive historical trends on
that moment was so real as to be almost tangible." Jones
recalled with satisfaction, "Group drafting usually leads
to the lowest common denominator of policy content.
In this case there were no cautious voices raised. All
concerned were agreed that the President should address Congress and the American people in bold policy
terms."

That was Truman's goal from the start. As he later
wrote, "I wanted no hedging in this speech. This was

America's answer to the surge of expansion of Communist tyranny. It had to be clear and free of hesitation or double talk."

There was not a moment to lose. As Acheson wrote later, "Greece was in the position of a semiconscious patient on the critical list whose relatives and physicians were discussing whether his life could be saved."

Just as he had upon the delivery of the British messages days before, Acheson took the lead in the administration's all-important approach to Congress. General Marshall had left State once more in the capable hands of the undersecretary. He ordered Acheson to proceed with the plans for Greece and Turkey with no regard for his negotiations with the Soviets; the Balkans were to take priority even if this led to failure of his mission to Moscow. An admiring Acheson would later marvel, "Many years would go by before an officer commanding in a forward and exposed spot would call down his own artillery fire upon his own position to block an enemy advance."

Acheson assigned the task of drafting the required legislation to Jack Hickerson, director of the Office of European Affairs.

Confronted with piles of paper—early drafts by Henderson and others, and the "Public Information

Program"—Acheson found himself pondering the best way forward. He admired Truman, and would later dedicate his famous memoir, *Present at the Creation*, to him, using a line from Edwin Markham's poem about Lincoln: "The captain with the mighty heart." But as he considered the task before him his mind drifted back to President Roosevelt, telling those around him, "If FDR were alive I think I know what he'd do. He would make a statement of global policy but confine his request for money right now to Greece and Turkey." Acheson was well aware that America was on the verge of a long-term and costly commitment to freedom throughout the world, and that many future requests would be made of Congress. But legislators needed to be eased into this brave new world, and it simply would not do to abruptly deliver an expansive wish list to Capitol Hill.

The discussion focused mainly on Greece. The situation there was more dire than in Turkey, and Acheson stressed that the speech would have to reflect that reality. The undersecretary read the papers spread before him and chose the most compelling passages that would focus the minds of Capitol Hill leaders on the democratic cause.

Curiously enough there were two words that would *not* appear in the speech: "Soviet Union." Nobody listening would have any doubt about the ultimate source of the threats against Greece, Turkey, and global security, but Acheson felt it would be needlessly provocative and diplomatically unwise to mention the Russians directly. Avoiding mention of the Soviets might also marginally shield the administration from accusations of saber-rattling from Republican isolationists and left-wing legislators still ambivalent about Stalin. The address was to be a declaration of support for Western democracy and individual freedoms, rather than one of open opposition to a particular foe.

There were other considerations the Truman administration had to take into account as well. Members of Congress and the general public might be moved by descriptions of suffering in the Balkans, or by a defense of popular democracy. But the Soviet alliance had ended just a year and a half before, and the American public was not as reluctant as the British people to abandon wartime notions of Stalin and his regime; but it would require more time and artful persuasion before most Americans would fully acknowledge the extent of the Soviet threat.

As would be the case over the next four decades, this Cold War the United States was about to enter would require difficult trade-off's in the fight to contain communism. Not for the last time, the United States would come to the aid of a repressive government in order to prevent a Soviet takeover. Truman would have to convince Congress and the American people not to make the perfect the enemy of the good. The fight to vanquish Hitler's Nazi tyranny was an unambiguously noble effort; coming to the aid of a Greek regime with thuggish right-wing characteristics was not.

While there were many cooks in the speechwriting kitchen, no one doubted that Acheson was the executive chef. He pushed back on those who thought the speech too provocative and resisted those who wanted an explicit rhetorical attack on the USSR. There was no dewy-eyed sentiment expressed toward the United Nations; international cooperation was all well and good, but could be dispensed with, the undersecretary believed, when reality required it.

Three days of effort went into the address, with Acheson and his lieutenants carefully weighing every word and its impact on their historic mission. Finally on March 6 a draft was ready for the secretary's consideration, and Acheson cabled it to Marshall in Paris, where the secretary of state had stopped en route to Moscow for a state dinner hosted by the president of France. The general quickly approved the text, with a few sharp edits, which were sent to the White House the next day for review. Marshall was keen to keep the process moving, but as one of his traveling aides was to put it, the general thought "there was a little too much flamboyant anti-Communism in that speech."

In the midst of the fast-paced drafting of the address, which was far from complete, the president gathered with his cabinet on March 7 to discuss Greece and Turkey and outline his expectations. Acheson provided his colleagues with an update, stressing that for the moment "Turkey is much better off"; "Greece is key to encirclement movements in France, Italy, Hungary, and Turkey."

Truman took Acheson's assessment in, paused, and then looked around at his cabinet. He brusquely told them that the United States was "going into European politics," and that he faced "the greatest selling job" of any president. His cabinet members unanimously affirmed their support of his decision and strategized how best to convince Congress and the public that once again becoming involved in a European conflict was a noble cause.

The man through whose hands the speech would pass next was one of the president's closest advisers, a tall, forty-year-old lawyer named Clark Clifford. Born on Christmas Day in Fort Scott, Kansas, he grew up in Missouri and received his undergraduate and law degrees from Washington College in St. Louis. Before the Second World War, Clifford was a successful and prosperous attorney. He was to make a life and career out of fortuitous connections; the first was a friend and fellow naval officer who was assigned to President Truman as naval adviser. Clifford went with him to the White House as an assistant, and then succeeded to the role. Truman found the young attorney so valuable that he named him White House counsel following his discharge from the navy.

Clifford cemented the relationship with Truman by organizing the president's beloved poker games on the presidential yacht. Thus he found himself in the company of the most powerful men in Washington for hours at a time, surrounded by cigar and cigarette smoke and privy to their personal and political secrets. Clifford also charmed the president's wife and mother, which further boosted him in Truman's esteem.

Thus he was launched on a glittering career that would eventually see him become a trusted adviser to four Democratic presidents, and a wealthy and distinguished Washington lawyer. During the Vietnam War, Clifford would step out from behind the scenes to serve Lyndon Johnson as secretary of defense following the resignation of Robert S. McNamara.

But that was in the future. For now, Clark Clifford had to take the State Department's draft and fashion it into a speech that Truman would be able to deliver with conviction, and that would convey to Congress and the American people the graveness of the situation. The raw material was in the draft, but Clifford's practiced eye saw ways to increase its dramatic impact.

The Washington lawyer's knowledge of the issues at hand was extensive, for along with his assistant George Elsey, he had crafted a top-secret, hundred-thousand-word report finished the previous September for an exclusive audience: the president of the United States and his top officials. In "American Relations with the Soviet Union," which would later become known as the Clifford-Elsey Report, the two former naval officers painted a bleak but accurate picture of the deteriorating relationship between the two wartime allies. If the Truman Doctrine was the public declaration of the Cold War, the Clifford-Elsey Report was its founding document. It contained echoes of Churchill's Iron Curtain speech and Kennan's Long Telegram, but the tone was even more urgent than the words contained in either of those documents. The focus of the report was almost exclusively ideological, with no attention paid to legitimate Russian security concerns that Kennan had so carefully illustrated. It warned that the Soviets were "on a course of aggrandizement designed to lead to eventual world

domination by the U.S.S.R." Across the globe, the Soviets were flexing their military strength, and even in
those countries where they had no military presence,
Stalin's regime was surreptitiously funding communist parties with a view to undermining those nations
from within. Fittingly enough—considering what was
to come—the report prominently featured Greece and
Turkey:

> The Soviet Union is interested in obtaining the
> withdrawal of British troops from Greece and the
> establishment of a "friendly" government there. It
> hopes to make Turkey a puppet state which could
> serve as a springboard for the domination of the
> Eastern Mediterranean. . . . The United States
> should support and assist all democratic countries
> which are in any way menaced or endangered by
> the U.S.S.R. . . . Providing military support in case
> of attack is a last resort: a more effective barrier to
> communism is strong economic support.

President Truman deemed the Clifford-Elsey report
so explosive that he ordered that every copy other than
his be put under "lock and key." Clifford, never bur-
dened by false modesty, considered his work to have

been highly influential, while Elsey, who had done the vast majority of the report's heavy lifting, was less certain.

Whatever the report's effect on Truman's mind, the speech that was now in Clifford's hands borrowed much from it. Though he was not a gifted writer, and would often depend on others to help express himself on paper, Clifford knew Truman's mind and had a sense of the dramatic. He suggested to Jones that the speech "build up steadily to a climax" by opening with a brief survey of conditions in Greece and Turkey, and then moving toward the "wider situation." Showing the political savvy that would benefit him throughout most of his career, Clifford also suggested the addition of a paragraph emphasizing that the Truman administration would closely supervise all foreign American expenditures. Finally Clifford advised that the speech conclude with a rousing call to action.

Jones made the suggested changes and returned the speech to Clifford, who spent Sunday the ninth with George Elsey, adding further revisions. In *Proclaiming the Truman Doctrine*, for which she interviewed Elsey many years later, Denise M. Bostdorff provides an account of how the two presidential aides set about to "Trumanize" the speech that day:

[They] quickly deleted excess verbiage. . . . Simple words also took the place of more complicated terms, so that "rectified" became "corrected," "expenditure of any funds" became "use of any funds," and "The government of Greece has its imperfections" became "the government of Greece is not perfect." . . . Clifford and Elsey also realized that simpler sentences were essential for the president's formal addresses because Truman's poor eyesight made it difficult for him to look down at a text, look up at the audience, and then refocus on the text. The president, Elsey reflected, "did not read well. His delivery was not that of a polished orator when he was reading a formal address. It was just a fact of life. We accepted it. This is what you were working with; you play the cards you're dealt. He did his damnedest, and we did ours."

Elsey was also concerned about the haste of their deliberations, writing in a memo to Clifford at the time, "I do not believe that this is the occasion for the 'all-out' speech," while lamenting that there was "insufficient time to prepare what would be the most significant speech in the President's administration. Much more time is necessary to develop the philosophy and ideas and do justice to the subject. I think the President should have two weeks to prepare such a speech." But the sweep of events rendered Elsey's objections moot.

On Monday the tenth of March, Clifford crossed West Executive Avenue with a copy of the address, as revised by him and Elsey and approved by the president. This would be the final opportunity for Acheson and his team to influence its contents before Truman went to the Hill two days later. The undersecretary recommended the removal of three new points that the White House staff had added:

A line that explicitly described Greece as a
   gateway to the Middle East;
A reference to Middle Eastern "resources";
   i.e., oil;
A warning that a growing adoption of centrally
   controlled economies around the world was
   threatening free enterprise at home.

According to Jones, Acheson was especially concerned with the third item, as Britain itself was undergoing a peaceful and democratic socialist revolution without endangering the freedom of its people. For Acheson, at issue was not the precise economic system chosen by a given nation, but whether it made that choice in a free and democratic way.

Earlier that same day, the president had gathered congressional leaders in the Oval Office for a final discussion of aid to Greece and Turkey before his speech, revealing his plan to ask for $250 million for Greece and $150 million for Turkey. This White House meeting with congressional leaders did not go as well as the first, belying a *New York Times* story on March 8 that described members of Congress "relieved that some concrete proposal was to be put before them to replace the vague forebodings that have been expressed to them by Administration spokesmen." Referring to "growing excitement" over the issue of Greece, it described the atmosphere on Capitol Hill as "generally favorable" to the president's likely proposals, with the expectation of a "full-scale debate on the foreign policy of the United States."

But in person, it was clear to Truman that opposition to his plans had grown as the state of the European economy had become clearer. Acheson recalled "a cool and silent reception." The new Republican majorities in the House and Senate were unwilling to grant the president a blank check, having been elected on a platform of small government and foreign policy retrenchment.

As the congressional leaders departed the West Wing, Senator Arthur Vandenberg spoke to the reporters assembled outside, praising Truman's "great candor" and announcing (somewhat inaccurately) that the president would "discuss the whole situation" before a joint session of Congress two days later, at noon.

In the *New York Times* the following day, reporter Harold B. Hinton noted that there was speculation on Capitol Hill that Truman "might extend the geographical scope of his advice to Congress to take in China, Korea, and even Italy," illustrating how Washington's rumor mill was running on all cylinders. But in a hopeful sign for the administration, Hinton wrote, "Most Republican leaders . . . inclined to the view that Congress would accede to the President's recommendations."

Hinton then reported on an "unusual" Republican conference that evening, during which Vandenberg told his colleagues—with even greater boldness than the administration—that aid to Greece "might prove to be symbolic of a general, round-the-world situation in which the same fundamentals were involved." He claimed, somewhat ingenuously, that he would await the president's address before making up his mind, but told his audience, "this is a matter which transcends politics. There is nothing partisan about it. This is national politics at the highest degree."

The statesmanship shown by a once-reluctant Vandenberg was encouraging news for the exhausted but exhilarated leadership manning the stations inside the White House and State Department. Still, political storm clouds remained over the White House. Even colleagues from Truman's party began feeling restive; a meeting of congressional Democrats—the details of which had leaked to the press—had come out strongly against supporting the regressive Greek regime. There would be no groundswell of enthusiasm for a new and generous program of aid. Truman had seen this showdown coming. He was now facing the greatest "selling job" of his political career. Only presidential leadership of the most vigorous character could marshal the necessary support to pull reluctant Democrats and Republicans to his side. The stakes for the upcoming speech were historic. Truman was as tough a political fighter as there was in Washington, but as his own staff members admitted, he was a poor orator. The substance of his words would have to carry the day more than the style of his delivery. History was now hanging in the balance.

Chapter 11
In This Fateful Hour

The president and his entourage filed through the Diplomatic Reception Room on the basement level of the White House, where Franklin Roosevelt had spoken to the American people through the darkest days of the Great Depression and Second World War with this fireside chats. Truman knew the challenge that he faced on this day was every bit as difficult as many that confronted his predecessor, but the current president was self-aware enough to know that he would be facing this challenge without FDR's winning charm. Exiting under the South Portico, the president was bathed in the sunlight of a warm spring day.

The bright skies were of little comfort. Truman carried with him not just the weight of his office, but the trepidation that comes from having to ask a hostile

GOP Congress for their support. Worse, he was feeling under the weather. With the remnants of a nasty cold sapping his energy, this momentous address would be even more difficult than usual for the inartful orator. He gathered his strength for the storm of attention that was sure to follow, and braced himself for the inevitable crush of congressional and press attention.

His daughter, Margaret—who was also feeling under the weather—followed behind to see him off. As an aspiring singer, she too was bracing herself to perform in front of a demanding audience, and said to one of her father's aides: "I know how Daddy feels." But doting daughter though she was, Margaret would obviously never know the pressures one felt when carrying the weight of the free world on one's shoulders. On a president's most mundane day, the burden is too great for most to contemplate. But in that historic moment, when his speech could shape the future of Europe, the weight Harry Truman had to be carrying must have been unimaginable.

Truman moved into his limousine with Admiral William Leahy, the White House chief of staff, and other assembled aides. The First Lady followed behind in a separate car, and another behind that one carried Clark Clifford, press secretary Charles Ross, and a gaggle of other presidential assistants. Driving up

Pennsylvania Avenue, it traced in reverse—and more slowly—the route taken by Truman on the day he succeeded Roosevelt nearly two years before.

On that terrible day in 1945, he had joined his friend Sam Rayburn, then Speaker of the House, and other colleagues at the "Board of Education," the convivial drinking society of senior politicians that met in the Speaker's hideaway office for cards and drinks. There the whiskey flowed freely as did the old politicians' stories. Only after Truman had poured himself a drink did Rayburn remember to pass along a message: White House press secretary Steve Early had called. Truman returned the call with drink in hand, only to hear Early tersely order him to come to the White House at once. Turning pale, Truman gasped, "Jesus Christ and General Jackson." Though Early provided no further details, Truman suspected the reason for the call.

At their first meeting after the previous summer his selection as the vice presidential nominee, Truman had been shocked by the president's condition. Later he told a friend, "I had no idea he was in such feeble condition. In pouring cream in his tea, he got more cream in the saucer than he did in the cup. His hands are shaking and he talks with considerable difficulty. . . . Physically, he's just going to pieces." Now, as Truman raced through the Capitol toward his waiting car and driver, he feared that the worst had happened.

As he raced through the Washington night, Harry S. Truman was now the president of the United States. He just didn't know it yet.

Upon his arrival at the White House he was taken upstairs to a waiting Eleanor Roosevelt, who told him the news that he had long dreaded: "Harry, the president is dead." Overcome with grief and understandable shock, he could only ask the former First Lady, "Is there anything I can do for you?" She replied, "What can *we* do for *you*? For you are the one in trouble now."

The man from Independence, Missouri, told reporters the next day, "I felt as though the moon and the stars and all the planets fell on me last night when I got the news."

Almost two years later, Harry Truman was driving to the Capitol to deliver his momentous speech on the same day that the Senate passed what would eventually become the Twenty-Second Amendment to the Constitution of the United States, restricting presidents to two terms in office. Still chafing after four defeats in presidential elections and more than a dozen years of Roosevelt in the Oval Office, Republicans were eager to ensure that no future president would prolong their tenure past the eight-year precedent set by George Washington. Senator Robert A. Taft of Ohio, son of a one-term president and one of Truman's chief antagonists, introduced the amendment in the Senate. A final version would be sent to the states days later, and in February 1951 the amendment would be ratified. Thus Truman was about to address a body focused on restricting—not expanding—the power and scope of future Democratic presidents. As the incumbent, he would be exempt under the terms of the amendment, but Truman was on notice that this Republican Congress resented their treatment while languishing fourteen long years in the minority and were in no mood to meekly submit to the president's demands. The accretion of executive power often required by the necessities of war would not be maintained in a postwar world without a struggle. Truman would later write of the Twenty-Second Amendment, "The Republican 80th Congress took a sort of revenge on Roosevelt's memory because he'd made a lot of those people look bad by comparison," and he considered it, "excepting only the Prohibition Amendment, the worst thing that's ever been attached to the Constitution."

In the House chamber, the restless crowd settled slowly into their seats. For those who had been sworn in little more than two months before, the pageantry of a joint session was a new experience. Then as now there were 435 members of the House of Representatives, but on occasions like this the members shared their chamber with senators, members of the cabinet, Supreme Court justices, and other dignitaries. Former members of Congress, availing themselves of their lifetime access to the House floor, were also angling for seats. The galleries above were packed with press and visitors, and even the steps were filled of spectators. When the First Lady appeared, few took notice, and she seemed lost in the crowd until rescued by an alert White House aide.

The president pro tempore of the Senate took a seat on the podium next to the Speaker of the House, Joseph Martin, the only Republican to serve as speaker from 1931 to 1995. That prized position would normally be taken by the vice president of the United States, in his constitutional capacity as president of the Senate, but since Truman's elevation in 1945, the office of the vice president had been vacant. Twenty years would pass before the Twenty-Second Amendment would allow the president to fill such a vacancy between elections with the approval of both houses of Congress. Thus it was that Senator Arthur Vandenberg, the man who more than any other held the fate of the Truman Doctrine in his hands, found himself watching its unveiling while seated directly behind the president. The president knew that without Vandenberg, his proposal would be dead on arrival.

The chamber had been a constant witness to history since its completion in 1857, after the House of Representatives had outgrown its original quarters in the older sectioin of the Capitol (now Statuary Hall). Presidents Wilson and Roosevelt had both appeared there to request declarations of war in 1917 and 1941, respectively. The vast, windowless room had not pleased everyone when it was first completed; Benjamin B. French, commissioner of public buildings, called it a "monstrous salon" and "a kind of cellar, where none of God's direct light or air can come in. . . . A piece of gaudy gingerbread work, that will, in the end, do no credit to anyone who has had anything to do with it." It was certainly elaborate, with a Victorian ceiling of iron and stained glass 139 feet long by 93 feet wide, and a marble speaker's rostrum, both nineteenth-century remnants that would be replaced soon after Truman's speech that night.

At one o'clock, the sergeant at arms appeared and in a booming voice announced, "The President of the United States!" The crowd rose to its feet and applauded as Truman made his way down the aisle, escorted by a committee of congressmen and senators and shaking hands on either side of him, a black folder tucked under his arm.

Due to the dramatic circumstances of his elevation, and the unsettled state of the world, Truman had already addressed five joint sessions before this evening. The man who ascended the dais that evening was no longer the machine politician from Missouri, clearly overwhelmed by the responsibilities of his office, but a more confident president determined to rally the public to his cause. After he greeted the Speaker and the president pro tempore, and as the applause died away, Truman began to speak. His tone was clipped and businesslike, a marked difference from the sonorous, patrician voice of his predecessor, but one to which Congress and the country had already grown accustomed. This room had been witness to rhetoric far more soaring that what Harry Truman could provide, but rarely had the stakes been more dramatic.

He began: "The gravity of the situation which confronts the world today necessitates my appearance before a joint session of the Congress." And in spare, economic prose, he swiftly got to the point: "Assistance is imperative if Greece is to survive as a free nation."

There were those who would obviously dispute the notion that Greece was free; its government was authoritarian and the means it employed to remain in power in the face of communist agression were far from inspiring.

But what could not be disputed was the fact the Nazis' occupation coupled with a rising communist insurgency had left that ancient land on the brink of ruin. "When forces of liberation entered Greece they found that the retreating Germans had destroyed virtually all the railways, roads, port facilities, communications, and merchant marine." He continued setting the bleak scene. "More than a thousand villages had been burned. Eighty-five percent of the children were tubercular. Livestock, poultry, and draft animals had almost disappeared. Inflation had wiped out all savings."

Amid "these tragic conditions," the president explained that "a militant minority, exploiting human want and misery, was able to create political chaos which, until now, has made economic recovery impossible."

Truman explained to the packed House chamber that the government in Greece was no match for the combined challenges facing it. To deal with postwar economic devastation was challenging enough, but to reconstruct a country while fending off a communist insurgency would be unmanageable for any strugggling country.

Throughout his remarks, Truman returned time and again to the theme of America as an indispensable nation on the world stage. "There is no other country to which democratic Greece can turn. No other nation is willing and able to provide the necessary support for a democratic Greek government." Regarding American aid to Turkey, he declared, "We are the only country able to provide that help." British aid to Greece would be stopping at the end of the month, and bluntly added that America's chief ally was "reducing or liquidating its commitments" around the world.

In an effort to reassure Congress that any aid provided would be spent only as intended, he declared, "It is of the utmost importance that we supervise the use of any funds made available to Greece in such a manner that each dollar spent will count toward making Greece self-supporting, and will help to build an economy in which a healthy democracy can flourish."

Applause rippled through the chamber with this promise of fiscal oversight, but many in the audience remained skeptical that a large infusion of cash would facilitate Greek recovery rather than lining corrupt politicians' pockets.

But what of the United Nations, the international body that was only recently founded with great fanfare? The president dismissed the possibility of UN assistance in words that would often find an echo in the future: "We have considered how the United Nations might assist in this crisis. But the situation is an urgent one requiring immediate action and the United Nations and its related organizations are not in a position to extend help of the kind that is required."

Truman did not paper over the more unsavory aspects of the government in Athens. "No government is perfect," he acknowledged. But "one of the chief virtues of a democracy, however, is that its defects are always visible and under democratic processes can be pointed out and corrected." Acknowledging that the Greek government had "made mistakes," he refused to condone its actions, condemning "extremist measures of the right or the left." And he noted that for all its flaws, the government represented "eighty-five per cent of the members of the Greek Parliament who were chosen in an election" the previous year.

Having acknowledged the failings of Greece's government, the president then turned to Turkey. Though "spared the disasters that have beset Greece," America still needed to support Turkey to maintain its "national integrity," which was "essential to the preservation of order in the Middle East."

Truman then moved to the central message of the

speech, the one that would spark the fiercest opposition, and alter America's 150-year history of remaining aloof from the travails of other countries except in the most dramatic instances of international conflict: "I believe that it must be the policy of the United States to support free peoples who are resisting attempted subjugation by armed minorities or by outside pressures." This sentence, one of the most sweeping and consequential in the era in which it was delivered by an American president, was greeted with silence.

To lose this struggle so soon after surviving the Second World War would be "an unspeakable tragedy," Truman declared, not only for Greece and Turkey, but for other nations as well. The United States could not turn its back on the free world "in this fateful hour," for its own national interests depended on peace and stability across the globe.

Having laid out his case in stark and urgent terms, the Democratic president made his request for $400 million for the two afflicted nations. Reminding the Republican majority that the United States had spent a staggering $341 billion in the past war, he called aid to Greece and Turkey "an investment in world freedom and world peace" that was "little more than 1 tenth of 1 per cent" of the price tag of World War II. "It is only common sense," said the president, "that we should safeguard this investment and make sure that it was not in vain."

Having attended to the practicalities, Truman then appealed to his audience's emotions: "The seeds of totalitarian regimes are nurtured by misery and want. They spread and grow in the evil soil of poverty and strife. They reach their full growth when the hope of a people for a better life has died. We must keep that hope alive."

His closing words were both a challenge to Congress and a recognition of its pivotal role in shaping the Truman Doctrine: "I am confident that the Congress will face these responsibilities squarely."

The applause was polite, as it usually is after a presidential address to Congress. But there was little reason to believe at the time that this display of customary courtesy would rise to political support. The tone and temper of Congress had changed dramatically since that fateful day in 1945 when Truman appeared before it for the first time as president. The country was restless after a long Democratic reign, and the new Republican majorities were enjoying the prerogatives that came with their unaccustomed power. And Harry Truman, elevated to the highest office in the land only because of the untimely death of Franklin Roosevelt, had presented to them a proposal as vast and sweeping as could be imagined. He exited the chamber, leaving surprise and consternation in his wake, and raced to the airport to board his waiting plane for a much-needed holiday in Key West.

Though they could not predict public reaction to the president's radical proposal, Truman, Acheson, and the capital's foreign policy establishment likely knew that President Truman had sought to alter America's role in the world over the course of his twenty-minute speech. He was determined that the United States would no longer retreat into the cocoon of isolationism, and that America's post–World War I contraction that led to the rise of Hitler's Germany would not be repeated. Now the United States would take its rightful place on the world stage, providing support to the cause of freedom around the world. As Truman's plane made its way toward Florida, world events hung in the balance.

Harry Truman was keenly aware of the old Washington adage that "The president proposes, the Congress disposes." He could ask Congress for a down payment on American leadership, but would Congress approve? The currents of isolationism still ran strong through American society, and there was no guarantee that a single speech could overturn more than a century of isolationist instincts. The president's declarations would now have to be reduced to legislation, and the resulting bills would have to make their way through a tortuous legislative process. The streets of Athens were aflame. The Soviets were circling their prey. Would the Republican Congress act in time?

Chapter 12
Moving Beyond Monroe

The president had made his case to Congress and the country. Now it was time for the Senate and House to begin the debate on altering America's future role in the world. Should the United States, having just emerged victorious in another costly European war, take up the cause of that chaotic continent yet again? The rising tide of communism in the postwar world did not afford America or its allies the luxury of celebrating the spoils of that victory. Millions of soldiers had returned home looking for jobs. The reconstruction of Germany and Japan was still ongoing, costing untold millions of dollars. Most Americans believed that winning two European wars in a single generation was more than should be required of American troops, and many impatiently wondered when the arsenal of democracy would finally enjoy the windfall of a peace dividend.

Given that we know today about Stalin's crimes against his own people, it is hard to imagine that as late as 1947, most Americans saw "Uncle Joe's" USSR as an ally. Since Germany's invasion of Russia in June 1941, the Soviet Union joined with the Allies and bore the brunt of the fighting. The United States had provided vast amounts of aid to Russia during the war, as America's economic juggernaut churned out thousands of tanks, planes, weapons, and other matériel that was shipped in convoys to Archangel and Murmansk. As the United States slowly prepared to join the fray in North Africa in late 1942, and Italy in 1943, it was the Soviet Union that met Nazi forces in the unspeakably brutal battles of Stalingrad, Moscow, and Kursk. Stalin was angered by what he considered America's cynical alliance with his country, allowing the Russians to shed rivers of blood against the Nazi war machine as the United States fought in large part through economic means. The Soviet's sacrifice also made some Americans reluctant to contemplate war against the Soviets. But in time, Stalin's aggressive moves in Europe would eventually move American opinion.

Perhaps the only man who could successfully move reluctant Republicans on Capitol Hill toward supporting the Doctrine was Senator Arthur Vandenberg, chairman of the Senate Foreign Relations Committee, and president pro tempore of the Senate. It was Vandenberg who, in that latter capacity, had sat on the dais behind Truman as the president delivered his speech on aid to Greece and Turkey. The Michigan senator was not opposed to the broad outlines of Truman's plan, but what concerned him most, as he looked out upon a sea of troubled faces in the House chamber, was whether reluctant Republican colleagues and progressive Democrats who too often held a sympathetic view toward Russia would agree. Though a converted internationalist, he could well understand the reservations of his fellow members, for he had once been a staunch isolationist as well.

The most vigorous Republican champion of the Truman Doctrine seemed to have much in common with the president. Both were midwesterners born in 1884 to doting mothers who possessed great ambitions for their sons. Both saw their educational opportunities limited by the financial failure of their fathers. And both inherited political creeds from their families.

But there were profound differences between the two leaders' background and outlook. Vandenberg carried with him an intense passion for journalism and politics throughout high school, and confidently expected a successful career in both fields to unfold. Aged only twenty-two, he became editor of the *Grand Rapids Herald*, throwing himself into the political scene and plotting his own future ascent, with a seat in the Senate as his goal. For all his youthful enthusiasm, Truman never seriously contemplated a career in politics until he had failed in countless other ventures. The seminal experience in his life was service as an artillery officer in the First World War, a conflict which Vandenberg fervently supported at the time but one in which he never fought. In fact, he later came to see American involvement in the war as a tragic foreign policy mistake driven by profit-hungry corporations. Truman's election to the Senate was hard-fought and surprised most observers; Vandenberg enjoyed the luxury of being appointed to fill a vacancy.

Their differences deepened in the 1930s and beyond. Truman was a staunch supporter of FDR and based his 1934 Senate campaign on devotion to the New Deal. Vandenberg was a virulent critic of Roosevelt, especially as FDR carefully moved the United States toward involvement in the Second World War. After the passage of the Lend-Lease bill, a massive program of aid to Britain, Vandenberg raged, "We have tossed Washington's Farewell Address into the discard. . . .

We have taken the first step upon a course from which we can never hereafter retreat."

But in most personal dealings, Vandenberg was courtly and magnanimous, while Truman could be cutting and more likely to hold a grudge. Unlike the plainspoken Truman, Vandenberg absorbed the old traditions and mannerisms of the Senate, delivering ponderous orations. Truman's biographer David McCullough painted a memorable portrait of Vandenberg: "Large and hearty, he had the mannerisms of a somewhat pompous stage senator—the cigar, the florid phrase, and more than a little vanity, carefully combing a few long strands of gray hair sideways over the top of his bald head. When making a point on the Senate floor, he favored the broad gesture, grandly flinging out one arm in a sweeping arc." Walter Trohan of the *Chicago Tribune* observed, "Politicians as a class are vain but he was vain beyond most of the tribe." Like many of his colleagues, but perhaps more than most, Vandenberg was said to be in love with the sound of his own voice. And as the United States became more supportive of the Allied cause, seeming to edge ever closer to joining the war, Vandenberg used that voice to pillory FDR and call for America to refrain from involvement in Europe's latest bloodbath.

He had undergone what Acheson archly dismissed as "a political transubstantiation" in the wake of the Japanese attack on Pearl Harbor, which Vandenberg later said "ended isolationism for any realist." Gradually, the future Republican chairman shed his limited worldview and embraced America's new roles as the "arsenal of democracy" and defender of world freedom.

In 1943, in characteristically self-congratulatory fashion, he portrayed himself as "hunting for a middle ground between those extremists at one end of the line who would cheerfully give America away and those extremists at the other end of the line who would attempt total isolation which has come to be an impossibility." His most famous utterance was pithier: "Politics stops at the water's edge," a motto that may sound impossibly idealistic in today's partisan climate but was a guiding principle for most politicians throughout the postwar era. He was keen to involve himself in the details of the war effort, believing that the Senate should wield nearly as much influence over its conduct as the commander in chief. While this did not always endear him to the administration, it augured well for those who wished to see the United States remain a force for good in the world once Hitler was vanquished and the time for rebuilding Europe began.

During the war, British philosopher and historian Isaiah Berlin penned a revealing portrait of Vandenberg for the Foreign Office:

A member of an old Dutch family and a respectable Mid Western Isolationist. A very adroit political manipulator, and expert parliamentarian and skillful debater. He has perennial presidential ambitions, and is grooming himself into a position of elder statesman. He is something of a snob, not at all Anglophobe, and is a fairly frequent visitor at the White House and the State Department. In common with the rest of his State delegation he votes against the Administration's foreign policies, but has nothing virulent in his constitution and is anxious to convey the impression of reasonableness and moderation. He denies that he is or ever was an Isolationist, and describes himself as a Nationalist ("like Mr. Churchill").

Vandenberg's most famous pronouncement of his new worldview took the form of a Senate speech delivered three months before V-E Day, in which he announced his profound conversion. It became known, at least among his admirers, as "the speech heard around the world":

I do not believe that any nation hereafter can immunize itself by its own exclusive action. Since Pearl Harbor, World War II has put the gory science of mass murder into new and sinister perspective. Our oceans have ceased to be moats which automatically protect our ramparts. Flesh and blood now compete unequally with winged steel. War has become an all-consuming juggernaut. If World War III arrives, it will open new laboratories of death too horrible to contemplate. I propose to do everything within my power to keep those laboratories closed for keeps.

There were cynics in Washington who attributed Vandenberg's new perspective at least in part to his affair with Mitzi Sims, whose wealthy husband, Harold Sims, was an attaché (and likely a spy) stationed at the British embassy in Washington. The *New York Times* correspondent Arthur Krock later wrote waspishly, "Vandenberg's romantic impulses led to gossip at Washington hen-parties, where the hens have teeth and the teeth are sharp, that Vandenberg had been 'converted' from isolationism by the pretty wife of a West European diplomat, a lady of whom, as the saying goes, he saw a lot." Trohan of the *Tribune,* a paper unsympathetic to Vandenberg's new internationalism, tartly referred to the Michigan legislator as the "Senator from Mitzi-Gan."

Whatever the cause of his conversion, the president now could expect cooperation from Vandenberg, however difficult the senator might sometimes prove to be. But a struggle ahead remained along partisan and ideological lines since Vandenberg was still a member in good standing of the party that wished to see federal expenditures drastically reduced.

Not long after the White House meeting of February 27, Vandenberg wrote to a colleague:

I am frank in saying that I do not know the answer to the latest Greek challenge because I do not know the facts. I am waiting for the facts before I say anything. . . . But I sense enough of the facts to realize that the problem in Greece cannot be isolated. . . . On the contrary, it is probably symbolic of the worldwide ideological clash between Eastern communism and Western democracy; and it may easily be the thing which requires us to make some very fateful and far-reaching decisions.

Dean Acheson was forever grateful for Vandenberg's support of the aid bill, but looked upon his legislative method with a mixture of amusement and condescension. As he put it, Vandenberg would "go through a period of public doubt and skepticism; then find a comparatively minor flaw in the proposal, pounce upon it, and make much of it; in due course propose a change, always the Vandenberg amendment. Then, and only then, could it be given to his followers as true doctrine worthy of all men to be received. . . . Its strength lay in the genuineness of each step. He was not engaged in strategy; rather he was a prophet pointing out to more earthbound rulers the errors and spiritual failings of their ways."

Despite those shortcomings, Vandenberg soon proclaimed his support for Truman's plan, declaring after the speech,

The President's message faces facts and so must Congress. The independence of Greece and Turkey must be preserved, not only for their own sakes but also in defense of peace and security for all of us. In such a critical moment the president's hands must be upheld. Any other course would be dangerously misunderstood. But Congress must carefully determine the methods and explore the details in so momentous a departure from our previous policies. . . .

The immediate problem may be treated by itself. But it is vitally important to frankly weigh it for the future. We are at odds with communism on many fronts. We should evolve a total policy. It must clearly avoid imperialism. It must primarily consult American welfare. It must keep faith with the pledges to the Charter of the United Nations to which we have all taken. . . .

We cannot fail to back up the President at such an hour. . . . We must proceed with calm but determined patience to deal with practical realities as they unfold. We must either take or surrender leadership.

Truman and his congressional supporters were fortunate that most of the press coverage of the president's speech had been, in the words of the government's Division of Press Intelligence, "exceedingly favorable," with opinion running "4 to 1" in favor of such action.

The *New York Times* opined after the speech that it would "launch the United States on a new and positive foreign policy of worldwide responsibility for the maintenance of peace and order." A *Washington Post* story called the president's address "one of the most momentous ever made by an American Chief Executive." An editorial in the *New York Herald Tribune* said that Truman "was asking for dollars; but he was also asking for the enthusiasm, the willingness to venture, the belief in our own values, which can prove to the shattered peoples of the world that the American system offers a working alternative to the totalitarian order which is otherwise their only refuge."

Of course, there were dissenters. The most strident among them was, to nobody's surprise, the communist *New York Daily Worker*, which lamented "a day of national shame for our country," and condemned "the empire-grab, masked by anti-Communist hysteria."

The phrase "Truman Doctrine" had appeared nowhere in the president's speech, and the State Department aides who had so hurriedly prepared for this moment had seen it as a new diplomatic stance and national policy. But the press soon christened it as such; possibly the first journalistic reference to it was in the *Washington Daily News* two days after the speech. "President Truman's message to Congress on relief to Greece and Turkey in effect is a corollary of the Monroe Doctrine. . . . While the implications of the 'Truman Doctrine' are as grave as any the people of the United States ever were called upon to face, they are no more than those to which they were committed by the doctrine of Monroe. . . . Both the Monroe and Truman Doctrines are thoroughly and peculiarly American."

President James Monroe had laid out would become known as the Monroe Doctrine on December 2, 1823, in his annual State of the Union address to Congress. As was the custom then, such messages were delivered to Congress in writing, and read by a clerk. Just as the Truman Doctrine was largely the work of Dean Acheson, so the Monroe Doctrine was fashioned by the then secretary of state, John Quincy Adams. It declared "as a principle in which the rights and interests of the United States are involved, that the American continents, by the free and independent condition which they have assumed and maintain, are henceforth not to be considered as subjects for future colonization by any European powers. . . . We should consider any attempt on their part to extend their system to any portion of this hemisphere as dangerous to our peace and safety."

Momentous though it was, the Monroe Doctrine was then largely aspirational. The United States lacked the economic and military might to enforce Monroe's vision. And it was also an outgrowth of American isolationism, a desire not only to steer clear of involvement in European affairs, but also to keep European powers at bay. But the Truman Doctrine was declared by the leader of the most powerful nation on earth, a giant among the ruins of a war-torn world, with more than sufficient means to back it up. And rather than declaring a desire to steer clear from the rest of the world, Truman's "Doctrine" was a blueprint for global leadership. Thus it was much more than an echo of the Monroe Doctrine—it was an emphatic successor.

The Senate chamber that was Vandenberg's stage and theater was smaller than that of the House, but was historically vast nonetheless. It had become the home of the Senate in 1859, leaving behind the old chamber that had echoed with the oratory of Clay, Webster, and Calhoun. The senate moved its operations the same time the House moved into its current quarters. Robert Caro has evocatively described the Senate space as it was in the late 1940s:

From its upper portion, from the galleries for citizens and journalists which rimmed it, it seemed even longer than it was, in part because it was so gloomy and dim—so dim . . . when lights had not yet been added for television and the only illumination came from the ceiling almost forty feet above the floor, that its far end faded away in shadows—and in part because it was so pallid and bare. Its drab tan damask walls, divided into panels by tall columns and pilasters and by seven sets of double doors, were unrelieved by even a single touch of color—no painting, no mural—or, seemingly, by any other ornament . . . . The only spots of brightness in the Chamber were the few tangled red and white stripes on the flag that hung limply from the pole on the presiding officer's dais, and the reflection of the ceiling lights on the tops of the ninety-six mahogany desks arranged in four long half circles around the well below the dais.

But the plain chamber possessed a grandeur nonetheless; it was apparent to those on the floor, who could appreciate the marble of the dais, "deep, dark red lushly veined with grays and greens, and set into it, almost invisible from the galleries, but, up close, richly glinting, were two bronze laurel wreaths, like the wreaths that the Senate of Rome bestowed on generals with whom it was pleased, when Rome ruled the known world—and the Senate ruled Rome."

The Senate may not have ruled the United States—it was, after all, just one of two chambers in Madison's constitutional system of checks and balances spread across three branches. But in this particular moment, the chamber would determine the fate of Truman's post-war vision. Vandenberg may have given his blessing in advance, but there were stormy seas to navigate before the bill to aid Greece and Turkey could be safely steered into port.

The chairman of the Foreign Relations Committee sought to avoid any legislative "rocks" that might appear on that journey by making the process as transparent as possible. Before introducing the Senate version of the bill, he summoned members of the administration to address the committee in private on the day after the president's speech. Many questions would have to be answered before the debate could be thrown open to the glare of public scrutiny.

In an unusual move, Vandenberg invited every senator to submit questions about aid to Greece and Turkey that would be referred to the State Department for detailed replies. In a measure of the importance attached to the issue, senators submitted four hundred questions, which the committee staff edited down to 111. The resulting questionnaire was delivered to State on March 20. The harried bureaucrats of the State Department, determined to expedite the process on Capitol Hill, returned detailed answers nine days later.

Senator Vandenberg formally began the deliberative process by introducing S. 938 on March 19, one week after Truman's speech. The six-page bill authorized the president to "from time to time when he deems it in the interest of the United States furnish assistance to Greece and Turkey, upon request of their governments, and upon terms and conditions determined by him." The legislation reserved the right of American government officials to confirm that funds provided were spent properly. The total authorization granted for the program was $400 million, and the president was given full authority to determine the proportion of aid for the two countries. It was a historically sweeping grant of power to the executive, recognizing the president's longstanding authority in the realm of foreign policy.

But the Senate Foreign Relations Committee would soon demand its own day in the sun.

On Monday, March 24, Senator Vandenberg gaveled the committee to order in Room 318 of the Senate Office Building (today the Russell Building). The committee had made its home in a suite of elegantly appointed rooms in the Capitol building since 1933, but they were small and intimate, suitable only for private hearings and hosting foreign visitors. For consideration of an issue of such overwhelming interest, the much larger hearing room across Constitution Avenue was necessary.

That room was full of spectators and reporters as Vandenberg began the hearing by announcing the American Legion's support for the bill, and inserting a statement by the Legion's national commander in the record. The Legion was founded in the wake of the First World War to represent American veterans, and with a million members it wielded immense influence in communities around the country. The statement urged that America be "world-minded in its states-manship," and "become the great rallying center for democracies everywhere in the struggle to preserve and to secure the freedom of all peoples."

The chairman then recognized the undersecretary of state, who would lay out the administration's case for the legislatioin. Acheson read at length from a prepared statement, again stressing publicly that aid to Greece was a necessary national security measure.

Acheson vividly described the devastation that years of war had inflicted on Greece, and the "severe strain" on the Turkish economy. He assured the members that the proposal did not involve sending troops to Greece or Turkey: "We have not been asked to do so. We do not foresee any need to do so. And we do not intend to do so." Only "observers and advisers" would be sent— true enough in this case, but words that would later echo ominously in the build up to the Vietnam War.

Responding to concerns that the Truman Doctrine was too sweeping, Acheson reassured the commit-tee that "any requests of foreign countries for aid will have to be considered according to the circumstances in each individual case." And to those who thought that the policy risked war with the Soviets, he said simply, "Quite the opposite is true." The promotion of stability and democracy would lead not to war, but "in the other direction."

During his statement, he quoted the American statesman Daniel Webster, who during his tenure in the House of Representatives spoke in support of aid to Greece in its struggle against the Ottoman Empire. Webster's words, spoken in 1823, seemed suited to the situation in 1947: "Mr. Chairman, there are some things which, to be well done, must be promptly done. If we even determine to do the thing that is now proposed, we may do it too late. . . . With suffering Greece, now is the crisis of her fate—her great, it may be, her last struggle. Sir, while we sit here deliberating, her destiny may be decided."

A number of committee members returned to the issue of the bill's broader implications. The ranking member, Senator Tom Connally of Texas, pressed Acheson further, saying, "This is not a pattern out of a tailor's shop to fit everybody in the world and every nation in the world, because the conditions in no two nations are identical. Is that not true?" Acheson agreed and repeated his earlier contention that any future requests for aid would "have to be judged . . . according to the circumstances of each specific case." "It cannot be assumed," he reiterated, "that this Government would necessarily undertake measures in any other country identical or even closely similar to those proposed for Greece and Turkey." He cynically rejected the notion that the aid bill was a "doctrine" in the spirit of Monroe's pronouncement of 1823. His caution was understandable, but given the president's sweeping rhetoric in his March 12 address to Congress and Acheson's central role in crafting it, the undersecretary of state was being more than a little disingenuous.

Chairman Vandenberg, the former isolationist with the zeal of the converted, expressed concern at Acheson's restrained interpretation: "A good deal of emphasis has been put this morning upon localizing this project in Greece and Turkey so far as precedent is concerned," he said with disapproval. He quoted the president's assertion that "totalitarian regimes imposed on free peoples . . . undermine the foundations of international peace and hence the security of the United States," and asked Acheson to "broaden the concept which we are discussing this morning."

Vandenberg's fervent support had the effect of putting Acheson in an awkward position. Under the glare of the press, in the first and most important Senate hearing on the matter, he was being encouraged by the administration's most important congressional ally to echo the president's sweeping tone. But many in both houses of Congress were concerned about granting the administration carte blanche to intervene in countries spread across the globe. The undersecretary responded cautiously by denying that the president intended to launch an "ideological crusade," and that while Truman had expressed a "broad principle" in favor of free institutions, the matter at hand was restricted to Greece and Turkey.

The chairman persisted: "I think what you are saying is that whenever we find free people having difficulty in the maintenance of free institutions, and difficulty in defending against aggressive movements that seek to impose upon them totalitarian regimes, we do not necessarily react in the same way each time, but we propose to react."

In his dramatic firsthand account of the birth of the Truman Doctrine, Joseph Jones highlighted the importance of Vandenberg's last five words, marveling that the former isolationist was "insisting upon, and himself restating, the global implications of the Truman Doctrine." There would be many times in the future when a committee chairperson would try to restrain the executive in the arena of foreign affairs, but in this instance a powerful senator was encouraging a secretary of state to be bolder and more assertive.

Still, the chairman was content with the administration's performance that day. Vandenberg appreciated Acheson's powerful testimony and tireless assistance, saying later, "I have never seen such willingness to cooperate with the legislature. I think if I called him at ten in the morning and asked him to deliver the Washington Monument to my office by noon he would somehow manage to treat this as a proper request and deliver it."

On April 28, it was Ambassador Lincoln MacVeagh's turn to appear before the committee in the second of eight executive sessions it held on the bill. The ambassador had just flown in from Athens before the hearing. No other American was as familiar with events on the ground in Greece as MacVeagh, and his diplomatic dispatches over the previous year had been enormously helpful in preparing his government for the trial to come. But in his testimony before the Senate, the ambassador displayed far less nuance and sophistication in his presentation. A good political soldier, he knew that a more direct approach was required, one that focused on the growing communist influence in Greece. In keeping with the administration's line, he painted a vivid picture of the ongoing subversion in the country, highlighting outside pressures. And given the private nature of the hearing, in the committee's gilded and intimate room in the Capitol, he could afford to be more freewheeling in his account. His warning to the committee was stark:

> At the present moment, the situation is exceedingly grave and critical, actually critical. Any delay . . . is very dangerous if we are going to avoid a total collapse of the country, both economically and socially, which will bring the country into the satellite orbit of the Russian Empire.

The ambassador made clear the harm that Greek society had already suffered, in terms more stark than he could have used in public, calling Greece "a country that is full of people who have not got any sense of responsibility. . . . It is an awful mess."

Questioned by Vandenberg about the difficulty of putting down the communist guerrillas, MacVeagh said he remained optimistic:

> I think that the organizers, these fellows who are organizing the unfortunates in Greece, and that is what it amounts to—the unfortunates and miserable have gone into the mountains and are organized

and tightened up and being formed into a weapon by an international Communist group. You break down their organization and you chase out or capture the fellows who are organizing them, and you will have a certain amount of banditry in Greece for a great many years, but it will not be an organized subversive political movement. It will be just fellows in the hills like Robin Hood, who occasionally come down and carry somebody off for ransom.

The breezy confidence and informal approach that had characterized his testimony before the committee were nowhere in evidence in the secret telegram he sent the Greek prime minister soon thereafter. The government's extreme right-wing nature and the harsh measures being employed from Athens continued to weigh on the minds of those tasked with deciding whether to send aid to Greece, and MacVeagh, in diplomatic but unmistakable terms, read him the riot act:

While every effort is being made here to secure implementation of the President's program for aid to Greece, public opinion is being constantly disturbed by reports of official toleration of rightist excesses and the application of security measures to non-subversive political opponents of the govt.

The impression created by these reports is that the President's program aims to assist a reactionary regime with all the earmarks of a police state, which is an idea unacceptable to the American people.

You will remember my concern over this matter expressed to you on numerous occasions. You will also remember your assurance to me that in my absence your policy would be in accord with the President's message. I would now respectfully emphasize again, but with a new urgency born of a critical moment, the advisability (1) of your Govt.'s giving some clear factual evidence of its political tolerance and broad national character by proceeding with equal vigilance and severity against all lawlessness whether of the right or left, and (2) of its giving its actions in this respect the fullest and most persistent publicity. That the Government of Greece is "fascist" in mind and action is the argument which is telling more potently than any other against the President's program and it can be effectively answered only by the observed conduct of that Govt. itself.

The bipartisan cooperation between Vandenberg and the administration was impressive by any measure, but this special relationship had its limits as well. Keen to maintain his position in the eyes of his Republican colleagues, the chairman clashed with Truman on matters both minor and major.

In *Harry and Arthur: Truman, Vandenberg, and the Partnership That Created the Free World*, Lawrence J. Haas tells the tale of a hapless Democratic National Committee official who asked his Republican counterpart to sign with him a statement "reaffirming both parties' support of America's foreign policy as outlined by President Truman with the concurrence of Arthur H. Vandenberg." This inartful lurch by the DNC was a step too far for Vandenberg, who, as Haas relates, responded angrily on the floor of the Senate:

> When bipartisan foreign policy gets into the rival hands of partisan national committees, it is in grave danger of losing its precious character. No matter how worthy the announced intentions, it can put foreign policy squarely into politics which, under such circumstances, are no longer calculated to stop at the water's edge. . . . Bipartisan foreign policy is not the result of political coercion but of nonpolitical conviction. I never have even pretended to speak for my party in my foreign policy activities.

An issue of more lasting impact was the United Nations. Judging by the written evidence, Vandenberg seemed of two minds on the issue, writing in the aftermath of the president's speech, "I am frank to say I think Greece could collapse fifty times before the United Nations itself could even hope to handle a situation of this nature," and at the same time declared, "The Administration made a colossal blunder in ignoring the UN."

Whatever Vandenberg's view, the UN's haplessness was already a growing political problem: the White House faced the awkward fact that the American people had quickly taken the United Nations to heart. President Roosevelt's vision of international cooperation found favor with a world weary of war, and the American people were happy to share the burdens of postwar leadership with others. Left-wingers and GOP isolationists in both houses were eager to seize upon the oversight, whether out of sincere regard for the UN or simply a desire to sabotage a policy with which they disagreed.

The United Nations Conference on International Organization had been held in San Francisco just two years prior, in 1945, at the end of the European war but before Japan was defeated. Representatives of fifty nations gathered in Northern California to formalize arrangements that had been agreed to at the Dumbarton Oaks conference in Washington the year before. Roosevelt had died less than two weeks before the San Francisco gathering.

Dumbarton Oaks had produced the United Nations Charter, the treaty by which every member state would be bound. Its preamble pledged that the "peoples of the United Nations" would, among other things, "save succeeding generations from the scourge of war, which twice in our lifetime has brought untold sorrow to mankind," and "employ international machinery for the promotion of the economic and social advancement of all peoples."

Given the administration's determination to act unilaterally in aiding Greece and Turkey, it is ironic that the closing remarks of this historic conference on collective security were delivered by none other than Harry Truman himself. On only his twelfth day as president, Truman flew to San Francisco and declared to the delegates: "You have created a great instrument for peace and security and human progress in the world. The world must now use it! If we fail to use it, we shall betray all those who have died in order that we might meet here in freedom and safety to create it." He went on to laud the Charter's "machinery of international cooperation which men and nations of good will can use to help correct economic and social causes for conflict."

Thus it was politically predictable that a Gallup poll published on March 27 reported that 56 percent of respondents opposed American aid to Greece and Turkey without consultation with the UN. In the weeks following Truman's address, the number of Americans who believed that the UN should take the lead continued to increase.

The administration did what it could to finesse the issue. The US representative to the UN, Warren R. Austin, released a statement on the day after the president's speech declaring, "Support of the freedom and independence of Greece and Turkey is essential" to collective security, and "prompt approval by Congress of the President's proposal would be new and effective action by the United States in supporting with all our strength our policy in the United Nations."

But to Joseph Jones, "Popular American support of the United Nations during its first two years of existence was sentimental and unrealistic, an escape from the responsibilities of our unique power." American policy makers likely knew that the UN would never relieve the United States of the burden of moral, economic, and military leadership. The talking shop soon to be based at Turtle Bay would always be viewed by the

DOD and state to be ill equipped to deal with "modern forms of aggression: infiltration, subversion, economic pressures, wars of nerves, aid to rebel groups," the very problems afflicting Greece at that time.

On March 25, James Reston wrote of the administration's dilemma in the *New York Times*, explaining that while legislators were nearly unanimous in their desire to protect the United Nations, they were "disagreeing violently" over the best way to achieve this. In order to set a precedent, should the United Nations be consulted regarding aid to Greece and Turkey, even though it was ill equipped to deal with the situation? Or would it be best to avoid involving the UN at all, so as not to highlight its weaknesses? Reston warned his readers, "In the course of arguing about how to help it, both sides are advertising its weaknesses and hurting the one thing they all agree they want to help."

The famed syndicated columnist Walter Lippmann, who later that year would introduce the term "Cold War" into the popular lexicon, supported the president's aims but also worried about the already diminished role of the UN:

It was, to put it conservatively, an oversight to have discussed our proposed action before the President announced it to the world only with the British government, and not also with the French, the Chinese, and the Russian. That, however, is water over the dam. . . . A full explanation, and a willingness to consider objections, would meet the obligation to consult. . . . We could inform the secretary-general of the United Nations about our proposals, and invite him to take notice of the explanations which will be offered to Congress. Mr. Austin could go before the Security Council and explain the proposals, not waiting until Mr. Gromyko attacks them. We could notify the United Nations that we shall not only explain what we intend to do and why, but also that we shall report to them at regular intervals what we have done and why. We could, moreover, invite the leading interested nations to send official observers to Greece to see for themselves what we are doing.

The great advantage of some such action on our part is that it would at once rehabilitate the moral authority of the United Nations and would reaffirm our loyalty as a member of the organization. It would not interfere with the efficiency of our action. . . . This is the best way to answer the charge that we are doing what we have so often charged others with doing—that we are acting unilaterally and for the purpose of domination and aggrandizement.

Vandenberg had been one of the chief congressional champions of the UN, and whatever his private estimation of its abilities, he wished to do nothing now that would undermine its international reputation. Besides, if the bill's opponents were successful in mobilizing pro-UN sentiment against it, disaster might follow. Thus for both personal and philosophical reasons, Vandenberg was keen to address the issue.

And if, as Acheson wryly observed, it allowed him to pose as "a prophet pointing out to more earthbound rulers the errors and spiritual failings of their ways," so much the better for him.

Thus, in consultation with Acheson and his House counterpart, Vandenberg introduced on March 21 an amendment to S. 938 that added a lofty preamble about the importance of the United Nations:

Whereas the Governments of Greece and Turkey have sought from the Government of the United States immediate financial and other assistance which is necessary for the maintenance of their national integrity and their survival as free nations; and . . .

Whereas the Security Council of the United Nations has recognized the seriousness of the unsettled conditions prevailing on the border between Greece on the one hand and Albania, Bulgaria, and Yugoslavia on the other, and, if the present emergency is met, may subsequently assume full responsibility for this phase of the problem as a result of the investigation which its commission is currently conducting; and . . .

Whereas the Food and Agriculture Organization mission for Greece recognized the necessity that Greece receive financial and economic assistance and recommended that Greece request such assistance from the appropriate agencies of the United Nations and from the Governments of the United States and the United Kingdom; and . . .

Whereas the United Nations is not now in a position to furnish to Greece and Turkey the financial and economic assistance which is immediately required; and . . .

Whereas the furnishing of such assistance to Greece and Turkey by the United States will contribute to the freedom and independence of all members of the United Nations in conformity with the principles and purposes of the Charter: Now, therefore be it . . .

This great senatorial gust of hot air was a genuflection before the UN, and gave its backers a fig leaf to help camouflage the uncomfortable reality that the United States was taking decisive and unilateral action.

So as to remove doubt among those who feared the nascent UN might be compromised by the Truman Doctrine, Vandenberg introduced another, more elaborate amendment on the following day. It read:

The President is directed to withdraw any and all aid authorized herein under any of the following circumstances:

If requested by the Governments of Greece or Turkey, respectively, representing a majority of the people of either such nation;

If the President is officially notified by the United Nations that the Security Council finds (with respect to which finding the United States waives the exercise of the veto) or that the General Assembly finds that such action taken or assistance furnished by the United Nations makes the contin-

uance of such assistance unnecessary or undesirable;

If the President finds that any purposes of the
Act have been substantially accomplished by the
action of any other intergovernmental organizations or finds that the purposes of the Act are incapable of satisfactory accomplishment.

This emollient language represented no risk to the
process—there was virtually no chance that either
Greece or Turkey would turn down millions in foreign
aid. Regarding the second point, the United Nations
was not consulted because the administration knew it
lacked the resources to act, a situation that was unlikely
to change in the near future. Finally, the third clause
gave the president the authority to determine whether
the objective of the aid had been achieved, a decision
Truman would be in no hurry to make. It was a clever
bit of legislative legerdemain that helped to defuse what
might otherwise have been a pressing problem for all
invested in the bill's passage.

The committee adopted both amendments by voice
vote—with no objections. It is fascinating to reflect,
seven decades later, how large the United Nations
loomed in the minds and imaginations of American
legislators. The Senate, which had rejected the League

of Nations in 1919, overwhelmingly approved the UN Charter in 1945 by a vote of 89 to 2. In the spring of 1947, the glow of its recent founding gave it a cachet that the international body would never enjoy again. It is impossible to imagine today Congress taking swift bipartisan action only after both parties were satisfied that the UN had been given its due respect and recognition.

Of the voices that were raised in opposition to Greece's rescue, perhaps the most influential was that of Henry A. Wallace, who had been Roosevelt's vice president from 1941 to 1945. Tall, urbane, and wealthy, Wallace was a passionate man of the left. Driven to distraction by his politics and eccentric persona, conservatives in the party pushed Wallace off the ticket at the 1944 Democratic National Convention. After winning a fourth term, Roosevelt appointed Wallace secretary of commerce.

Wallace remained at Commerce after Roosevelt's death, as Truman wished to steady the public's nerves by maintaining stability in the cabinet. But as tensions with the Soviets rose, Wallace's outspoken criticism of American policy became too much for Truman. In September 1946 he publicly declared, "We have no more business in the political affairs of Eastern Europe than Russia has in the political affairs in Latin Amer-

ica, Western Europe, and the United States." Such geopolitical declarations were for the commander in chief
to pronounce, not the secretary of commerce, and Truman wisely sacked him.

Even out of power, Wallace retained influence. As
editor of the prestigious liberal opinion journal *New
Republic*, and a tireless campaigner, FDR's former vice
president remained in the public eye, championing
left-wing causes.

He wasted no time in attacking the Truman Doctrine, taking to the airwaves on the day after the president's speech, saying, "President Truman calls for
action to combat a crisis. What is this crisis that necessitates Truman going to Capitol Hill as though a Pearl
Harbor has suddenly hit us? How many more of these
Pearl Harbors will there be? How can they be foreseen?
What will they cost?"

Even Churchill did not escape his wrath: "One year
ago at Fulton, Missouri, Winston Churchill called for
a diplomatic offensive against Soviet Russia. By sanc-
tioning that speech, Truman committed us to a policy
of combating Russia with British sources. That policy
proved to be so bankrupt that Britain can no longer
maintain it. Now President Truman proposes we take
over Britain's hopeless task. . . . I say that this policy is
utterly futile."

Fortunately for Truman, Wallace followed this broadside with an ill-considered lecture tour of the United Kingdom, during which he criticized the "ruthless imperialism" of the American government, declaring, "I shall go on speaking for peace, wherever men will listen to me, until the end of my days." Attacking his own country on foreign soil did Wallace no favors with the American public or press, and made opposition to the Truman Doctrine a political position outside the mainstream of American political thought.

Truman ignored Wallace's fulminations, but Churchill was rightly incandescent. In a speech at the Albert Hall in London for the Primrose League, an organization founded in memory of the great nineteenth-century British prime minister Benjamin Disraeli, he retorted:

We have had here lately a visitor from the United States who has foregathered with that happily small minority of crypto-Communists who are making a dead set at the foreign policy which Mr. Ernest Bevin, our Foreign Secretary, has patiently and steadfastly pursued with the support of nine-tenths of the House of Commons. The object of these demonstrations has been to separate Great Britain from the United States and weave her into the vast system of Communist intrigue which radiates from Moscow. Now I travel about a certain amount myself, and am received by much kindness by all classes, both in Europe and America. But when I am abroad I always make it a rule never to criticise or attack the Government of my own country. I make up for lost time when I come home. But when I am abroad and speaking to foreigners I have even defended our present Socialist rulers, and always I have spoken with confidence of the future destiny of our country. Here at home we must do our duty, point out the dangers, and endeavour so to guide the nation as to avoid an overwhelming collapse. But I have no patience with Englishmen who use the hospitality of a friendly nation to decry their own. I think this is a very good principle, and one which deserved general attention. . . . This is particularly appropriate where foreign policy is concerned.

Those who doubt the importance of an individual dramatically impacting the course of world events should ponder this: Had Wallace not been pushed out at the 1944 Chicago convention, or had Truman refused the president's invitation to join the ticket, then it might have been Wallace who succeeded Roosevelt upon the latter's death. The United States would have entered the postwar world with a socialist in the White House, with incalculable consequences for decades to follow.

A ferocious skeptic of big government, Ohio senator Robert Taft had been one of Roosevelt's chief congressional antagonists, and he continued his opposition to the domestic ambitions of the Democrats after Truman arrived in the White House. Known as "Mr. Republican," the Ohio senator stood for an uncompromising brand of small-government conservatism and used his mastery of parliamentary procedure to thwart Truman's liberal domestic ambitions.

Taft's father had served as President Theodore Roosevelt's secretary of war, and was Roosevelt's handpicked successor when the Rough Rider decided not to run again for the presidency in 1908. The elder Taft had little interest in the White House; his fondest ambition was to become chief justice of the United States. But he acceded to Roosevelt's wishes (and those of his wife) and duly won the election, helped by the popularity of his predecessor. President Taft found the job a miserable experience, and Roosevelt, his ambition unquenched, turned against his former protégé and resolved to run again in 1912. Failing to wrest the Republican nomination from Taft, TR ran instead as the candidate of the Progressive Party. The resulting split in the Republican vote gave the Democratic candidate, Governor Woodrow Wilson of New Jersey, the White House. Taft was stung by his defeat, but later achieved his ultimate ambition when President Warren G. Harding appointed him chief justice in 1921.

The younger Taft, undeterred by his father's past political travails, resolved from an early age to reach the White House himself. A brilliant student, he graduated at the top of his class from the preparatory school founded by his uncle, and performed similar academic feats at Yale and Harvard Law School. He established himself as a successful attorney and soon founded his own law firm. But all this was mere preparation for what he considered his ultimate calling. Following in the footsteps of his famous father, Taft entered politics and was elected to the Ohio legislature in 1920, serving for a decade and rising to become Speaker. In 1930 he was elected to the state senate, and although defeated for reelection two years later, he still burned with ambition

He won a seat in the United States Senate in 1938, and soon Senator Taft established himself as a leader of the conservative opposition to the New Deal. As the clouds of war gathered over Europe at the outset of his time in the Senate, the Ohio conservative was outspoken in his opposition to Roosevelt's pro-Allied policies, warning against American involvement in what was sure to be a devastating conflict. In 1940 he was a candidate for the Republican nomination for president, but with war seemingly imminent, his rock-ribbed isolationism was then out of favor even with Republicans and he lost the GOP nomination to political neophyte and internationalist Wendell Willkie.

Taft's commitment to isolationism was far from being the outgrowth of a sheltered existence. As a boy, he lived in Manila for four years while his father served as governor of the Philippines, an American colonial acquisition following the Spanish-American War. Later he would serve in the US Food Administration in the First World War, and was posted to Paris as a legal adviser to the American Relief Administration in the wake of the 1919 armistice. Taft was responsible for the coordination and distribution of humanitarian aid, and saw firsthand the devastation that four years of war had wreaked upon Europe. The experience impacted the future senator greatly and made him resolve to keep the United States far removed from the affairs of that troubled continent.

Those insights shaped his political identity, and therefore it surprised few that he responded coldly to Truman's appeal, telling reporters, "I do not want war with Russia. Whether our intervention in Greece tends to make such a war more probable or less probable depends upon many circumstances regarding which I am not yet fully advised and, therefore, I do not care to make a decision at the present time."

The *New York Times* reported Taft's view of the Truman Doctrine as an acceptance of "the policy of dividing the world into zones of political influence, Communist and anti-Communist," and Taft worried that "if we assume a special position in Greece and Turkey, we can hardly longer reasonably object to the Russians continuing their domination in Poland, Yugoslavia, Romania, and Bulgaria."

The Republican independence of mind and idiosyncratic approach to politics made it difficult to predict whether he would end up supporting the legislation, but his influence within the party and across America made winning him over an important goal for Truman.

The swift and dramatic transition from an internal administration process—one that, while discussed in exhaustive detail, had been almost entirely free of serious debate and discord—to a public debate open to endless scrutiny predictably raised concerns about the Truman Doctrine that had not been seriously vetted by its architects.

One such question involved the role of Turkey. The crisis in Greece had dominated that initial flurry of meetings and discussions in the State Department, and as Truman's speech had gone through the drafting process both there and in the White House, Turkey had taken a back seat in deliberations. Now that the matter was open for public debate, it was reasonable for skeptics of an expansive US role in world affairs to ask whether the inclusion of Turkey in the bill was an overreach.

One witness before the House committee put the matter somewhat caustically when he declared, "Administration and congressional spokesmen have themselves volunteered only the briefest discussion of the Turkish problem or of proposed United States aid to Turkey. Now, I do not mean to be ribald, but I really cannot help saying this: It almost appears that when the new dish was being prepared for American consumption Turkey was slipped into the oven with Greece because that seemed to be the surest way to cook a tough bird."

A tough bird Turkey may have been, but its rhetorical neglect in Truman's message was not meant to imply that it was any less strategically important. The situation in Greece was simply more urgent; by all accounts, the country itself was on the cusp of collapse and communist guerrillas had been successful in destabilizing the nation while exhausting the will of Greece's resistance. Turkey had shown itself more capable of fending off communist aggression and therefore found itself in a more stabile state of affairs.

There was yet another reason to keep the spotlight primarily on Greece. While the Soviets were prepared to exploit every opportunity that arose before them, their ability to directly impact military events on the ground in Greece was less so than in Turkey. And while both countries were important to the Soviet's strategic designs, Turkey was more pivotal in the short term, due to the importance of the straits and its access to the Mediterranean. Thus the possibility of provoking a Soviet response was greater in the case of Turkey, which the administration, for all its grim resolve still hoped to avoid. The less said about this Soviet target, the better.

But the issue of Turkey now had to be forthrightly addressed, and the task would fall to Vandenberg, who performed his job ably in his usual orotund way. The formal Senate debate on S. 938 began on April 8, after many hearings in the Senate and in the House. As the bill's sponsor and chairman of the Foreign Affairs Committee, Vandenberg was the first to speak, calling the proposal a plan to forestall aggression which, once rolling, could snowball into global danger of vast design. It is a plan for peace. It is a plan to sterilize the seeds of war. We do not escape war by running away from it. No one ran away from war at Munich. We avoid war by facing facts. This plan faces facts. . . .

If the Greeks, in their extremity, are not successfully helped to help themselves to maintain their own healthy right to self-determination, another Communist dictatorship will rise at this key point in world geography. Then Turkey, long mobilized against a Communist war of nerves, faces neighboring jeopardy. The two situations are inseparable. Turkey confronts no such internal extremity as does Greece; but it requires assistance to bulwark its national security. The president says that the maintenance of its national integrity is essential to the preservation of order in the Middle East. If the Middle East falls within the orbit of aggressive Communist expansion, the repercussions will echo from the Dardanelles to the China Sea and westward to the rims of the Atlantic. Indeed, the Middle East, in this foreshortened world, is not far enough away for safety from our own New York or Detroit or Chicago or San Francisco.

By explicitly drawing the connection between Greece and Turkey, and then the Middle East, Vandenberg made explicit what the administration had chosen to make implicit. Senators traditionally have had wider latitude to make certain pronouncements than presidents. Vandenberg warned grimly that the defeat of the measure "would be the forfeiture of all hope to effectively influence the attitude of other nations in our peaceful pursuit of international righteousness from now on. It would stunt our moral authority and mute our voice. . . . It would invite provocative misunderstandings of the tenacity with which we are prepared to defend our fundamental ideals."

Vandenberg's speech was the first of seventeen in favor of S. 938, and set the tone for all those to follow. The other speakers in favor were divided equally between Democrats and Republicans, an encouraging sign for the administration.

In a surprising development, Senator Taft ultimately voted in favor of the aid bill. His statement of reluctant support may have seemed to be an exercise in wishful thinking, but it was an act of responsible statesmanship by the conservative icon:

> I intend to vote for the Greek and Turkish loans for the reason that the President's announcements have committed the United States to this policy in the eyes of the world, and to repudiate it now would destroy his prestige in the negotiations with the Russian Government, on the success of which ultimate peace depends. I do not regard this as a commitment to any similar policy in any other section of the world. . . .
>
> I am in thorough accord with the Vandenberg amendments proposing that we withdraw whenever a Government representing a majority of the people requests us to do so, and whenever the United Nations find that action taken or assistance furnished by them makes the continuance of our assistance undesirable. I believe we should, in any event, withdraw as soon as normal economic conditions are restored.

Despite these stated reservations, Robert Taft surely knew that his dramatic decision to support the Truman Doctrine would help launch a worldwide crusade against communism.

As the Senate was in the midst of its grave deliberations, President Truman delivered another speech
in support of his doctrine, this one in more partisan
political surroundings. His motorcade left the White
House on the evening of April 5 and made the short
drive up Connecticut Avenue to the Mayflower Hotel,
a grand Washington institution that had opened two
decades before. It had already been the scene of several
notable events and presidential visits. Truman was here
on this night to address the annual Jefferson Day Dinner, sponsored by the Democratic National Committee
and held in honor of the third president—the party's
founder. His wife and daughter joined him on the dais,
as did Secretary of the Treasury John W. Snyder and
Congressman Sam Rayburn, who had lost his Speaker's
gavel after the 1946 midterm elections and now was
stuck with the unaccustomed title of House Minority
Leader.

Truman, gazing out at the party faithful gathered in
the vast ballroom, opened his remarks by quoting a letter from Jefferson to President James Monroe: "Nor is
the occasion to be slighted which this proposition offers
of declaring our protest against the atrocious violations
of the rights of nations by the interference of any one in
the internal affairs of another."

Monroe would later proclaim similar sentiments in the message to Congress that contained the doctrine that bears his name. Truman explicitly sought to ground his own recently proclaimed doctrine in the sentiments expressed by Monroe's, declaring:

We . . . have witnessed atrocious violations of the rights of nations. . . .

We too have declared our protest.

We must make that protest effective by aiding those peoples whose freedoms are endangered by foreign pressures.

The thirty-third president continued, moving far beyond any sentiments ever contemplated by Jefferson or Monroe:

We must take a positive stand. It is no longer enough merely to say "we don't want war." We must act in time—ahead of time—to stamp out the smoldering beginnings of any conflict that may threaten to spread over the world.

We know how the fire starts. We have seen it before—aggression by the strong against the weak, openly by the use of armed force and secretly by infiltration. We know how the fire spreads. And we know how it ends.

Let us not underestimate the task before us. The burden of our responsibility today is greater, even considering the size and resources of our expanded nation, than it was in the time of Jefferson and Monroe. For the peril to man's freedom that existed then exists now on a much smaller earth—an earth whose broad oceans have shrunk and whose natural protections have been taken away by new weapons of destruction. . . .

We are a people who not only cherish freedom and defend it, if need be with our lives, but who also recognize the right of other men and other nations to share it.

Truman concluded with one of the most stirring declarations ever written by Thomas Jefferson, one that he composed on the eve of his election as president and are still engraved on his neoclassical memorial on the Tidal Basin in Washington:

I have sworn, upon the altar of God, eternal hostility against every form of tyranny over the mind of man.

April 22 was the climactic day of the Truman Doctrine's debate in the Senate. The legislative day began at eleven a.m. with a prayer by the chaplain, Peter Marshall, who concluded with a plea to the Almighty, "Reveal to us now Thy word for today." Just as God was to reveal his word that day, so the United States Senate would reveal its judgment on S. 938.

This was the day chosen to close debate and vote on the five amendments proposed by Colorado senator Edwin C. Johnson before moving on to a final vote on the bill. Aware that his amendments would fail, and determined to make a last stand against the Truman Doctrine, Johnson withdrew them and introduced a substitute amendment that would eliminate all military aid from the package. Upon being recognized by the chair, he rose to speak and delivered a stinging attack on what he called a "pending war measure." Senator Johnson continued, eyes blazing,

I have ever been an implacable foe of peacetime conscription. . . . I have fought this Prussian-inspired military system at every turn of the road. . . . But if Congress enacts this measure pending war I will support conscription and military training, and a renewal of selective service on a wartime scale, immediate mobilization of a huge army and navy, increased military appropriations and every other step necessary to defeat our enemies quickly. . . .

Violently opposed to conscription and militarism as I am, this is an especially hard reversal for me to make, but our military alliance with Turkey which we are about to implement . . . leaves me no honest alternative. . . . I do not believe we can fight communism successfully with arms, but if that be our decision we must fight it in Moscow, not in Greece, not Turkey, not France, not all over the globe.

"Russia," he continued, "in her stupid, stubborn, exasperating policy of suspicious negation, and the United States, in her new dynamic policy of unilateral military intervention everywhere, are both dead wrong. . . . [The] UN must realize that time is running out and this is its last and best chance to avert World War III."

With tempers running high, Vandenberg denounced Johnson's statement as an "invitation to alien misunderstanding which otherwise would have no possible basis whatever."

As tempers cooled, the Senate began moving toward the vote. Senators began voting at four p.m., and Johnson's amendment was defeated easily, 68 to 22. A few desultory final amendments were then swatted away by the Senate leadership. Thousands of words had been spoken in support of and in opposition to S. 938 before the presiding officer finally called for a vote on the bill. The roll call began, with the clerk slowly reading each senator's name. When it was finished, the Truman Doctrine had taken a decisive step toward becoming a reality. S. 938 passed by an impressive margin of 67 to 23, with 35 Republicans and 32 Democrats in support, 16 Republicans and 7 Democrats against. The outcome was a triumph in bipartisanship.

Upon hearing the news, the Greek prime minister, Demetrios Maximos, told his parliament: "We assure President Truman that every dollar allotted to Greece will be appreciated as a symbol of the supreme industriousness of the American people and will be used exclusively for the purposes for which it was intended."

The passage of the Senate bill was not only a step toward ultimate congressional approval, but also a step toward the increased involvement of the United States in the political affairs of a foreign country, one with a painful past and a troubled present.

The *New York Times* reported the dramatic Senate vote and listed the yeas and nays. A front-page column the famed foreign correspondent Anne O'Hare

McCormick also acknowledged at length the weaknesses of the Greek government: "It apparently has no program but anti-communism and no capacity either to improve the lot of the people or to unite them by a generous gesture of reconciliation." Despite those failings, the legendary *Times* correspondent acknowledged that the political scene "is more like democracy than anything in the neighborhood."

McCormick then raised the critical question: Would the United States aggressively intervene in Greek politics if the House decided that America's new role in the world was to shield that country from "outside pressures"? Unfolding events in the region and across the globe would soon teach Harry Truman and his cabinet that there were few simple answers available to them in the troubled postwar world they had inherited from FDR.

Chapter 13
A House Divided

When I represented the first Congressional District of Florida in the US House of Representatives, my fellow members referred to our legislative body as "The People's House"; the Senate was scornfully dismissed as "The House of Lords." As members of the House, we understood that our constitutionally mandated role was to be the chamber most aligned with the current mood of Americans. House members are elected to two-year terms and senators serve for six years. Each House member represents about 600,000 constituents while Senators must cover an entire state. That is why the framers of the Constitution viewed the House to be the most democratic body in the federal government; it can also be the most raucous, with 435 members crowding into one chamber.

While there is no shortage of ego in the lower body, the sheer mass of humanity on the House floor tends to check the pretensions of most individual members. Besides, the Rules Committee leaders always made clear to rebellious members like myself that regardless of any feelings about the "People's House," it functioned legislatively more like a dictatorship than a democracy. The Rules Committee would set the limits of every debate taken up by the House. Among other things, this tightly limited the time members were allowed to discuss legislation and prevented the sort of long-winded orations that were so common in the Senate.

Though Dean Acheson could be a man of great charm, he did not suffer fools gladly, and on occasion he found it difficult to deal with House members, whom he considered provincial and ill informed. In his memoirs, Acheson noted an observation by historian Henry Adams that "the chief concern of the Secretary of State—the world beyond our boundaries—was to most members of the Congress only a troublesome intrusion into their chief interest—the internal affairs of the country, and especially of the particular parts of it they represented."

Truman's undersecretary of state ruefully reflected, "The principal consequence of foreign impact upon particular districts is trouble; rarely is it, or is it seen to be, beneficial. . . . The focus and representation of members of the House of Representatives are . . . narrowly circumscribed. For the most part, then, the Secretary of State comes to Congress bearing word of troubles about which Congress does not want to hear."

It was fortuitous, then, that the administration had as an ally the chairman of the House Committee on Foreign Affairs, Congressman Charles A. Eaton of New Jersey, a Republican Baptist minister accustomed to rallying followers to a cause.

"Doc" Eaton was in his twenty-second year in Congress when Truman made his momentous request on behalf of Greece and Turkey. The chairman was born in Nova Scotia in 1868, and moved to the United States to study theology in Newton Centre, Massachusetts. After his ordination he served as pastor of churches in Massachusetts, Ohio, New York, and Canada before settling in New Jersey and becoming a dairy farmer. A passionate preacher, his interest soon turned to politics, and he was elected to the House of Representatives in 1924. With his shock of white hair and neat mustache, Eaton was a striking and charismatic figure.

Chairman Eaton opposed the New Deal but was an ardent internationalist and anticommunist. Even before coming to Congress he was delivering speeches warning that bolshevism was "contrary to human nature" and that communists used "physical force as the instrument of its advancement." Along with Vandenberg, the Foreign Affairs Chairman in the House resolved immediately to make the Truman Doctrine a legislative reality. For all their political differences, Truman respected Eaton, finding him "enlightened" while praising his "brilliant, intelligent leadership." Those skills would be badly needed to pass Truman's sweeping legislation in an otherwise hostile Republican House.

Eaton introduced H.R. 2616, the House version of the bill, on March 18, a day before Vandenberg introduced the Senate version. In his floor statement, he swiftly scheduled a series of committee hearings, the first of which was on March 20.

The House chairman gaveled the March 20 hearing to order at ten a.m. Sitting before him in the flickering storm of photographers' flashbulbs was the familiar figure of the acting secretary of state. Acheson was prepared to deliver an even more "troublesome intrusion" than usual. The Groton and Yale man's discomfort on the House side of the Capitol would soon become apparent, even as he submitted to what he would describe to the committee members as an "ordeal."

Acheson immediately launched into a detailed account of the delivery of the British notes to the State

Department on February 21 and their immediate fall-
out. After the undersecretary's prepared statement had
been read, Eaton began to call upon each committee
member in turn, and a lengthy interrogation ensued.

The undersecretary of state engaged in a testy exchange with Congressman James P. Richards of South
Carolina, who asked a question about the UN and then
quoted from the Soviet newspaper *Pravda*, which had
asked the rhetorical question "Why did the United
States not bring the matter to the United Nations if
something really threatened Greece's integrity?" Acheson responded that he had already discussed the issue
in his opening statement, and asked irritably, "Is it
your desire that I repeat it?" Richards shot back, "I
wish you would amplify and clarify what you have said
in your written statement." The acting secretary responded with typical Achesonian arrogance: "Perhaps
saying something two or three times does clarify it."
The chairman likely winced; a witness talking down to
a member of Congress was bad form and could make
Eaton's job as Republican chairman even more difficult.

In response to a member who expressed concern
that "not only will we go after this procedure in Greece
and Turkey, but other nations threatened as they are
threatened," Acheson responded reassuringly, "You

must approach each situation as it occurs, or if it occurs, in the light of the facts of that situation. . . . We keep the field open to continue our choice."

Things took a more sober turn under questioning by Democratic congressman Mike Mansfield of Montana, who would go on to become Senate majority leader and ambassador to Japan. Mansfield's inquiry was blunt and direct: "In your opinion, what are the possibilities of this policy leading to war?" Acheson seemed momentarily taken aback before replying, "I should say that—I was going to say there was no possibility of it leading to war. I do not see how it possibly can lead to war. It seems to me that by strengthening the forces of democracy and individual freedom and strengthening the economies of those two countries, you do a great deal to eliminate the sort of situation which would produce friction between the great powers."

A future witness before the committee, Undersecretary of State for Economic Affairs William L. Clayton, would address the financial elements of the aid package in greater detail. President Truman had just created his post, and Clayton brought to it a wealth of experience, having served in senior positions in the Reconstruction Finance Corporation, the Export-Import Bank, and various government agencies. His statement to the committee described in detail the "appalling" state of the Greek economy in the wake of the war. Not only had the Germans ruthlessly exploited the country during the brutal occupation from 1941 to 1944, exporting food to Germany in vast quantities, but they had also wreaked havoc during their withdrawal. This "policy of systematic destruction" was "calculated to wreck the Greek economy to such an extent that a liberated Greece would have slight prospect of normal recovery in the foreseeable future." He continued, "The physical damage inflicted on the country was sufficient to result in almost complete paralysis. Means of communication were destroyed, port facilities wrecked, and bridges demolished. Livestock was carried off, villages burned, railways torn up, and the Corinth Canal dynamited."

Clayton concluded his statement by clarifying to members of the House Foreign Affairs Committee whether the aid to Greece would take the form of a loan or a grant. Some journalists had been unclear in their reporting, and even some of the committee members were unsure of the administration's intentions. The undersecretary emphasized that any American aid should take the form of outright grants, declaring, "I do not believe that we should create financial obligations for which there is no reasonable prospect of repayment."

Eaton's committee held a total of seventeen hearings on H.R. 2616, both public and private, and heard from a vast array of witnesses. The bill that gradually took shape during the deliberations differed slightly from the Senate version: it lacked Vandenberg's preamble about the United Nations and put greater emphasis on the need for "full and continuous publicity" about how the funds would be dispersed in Greece and Turkey. The committee finally reported H.R. 2616 on April 25 with only one dissenting vote.

The House floor debate on the momentous bill began on May 6, and was from the beginning a more contentious process than it had been in the Senate. Chairman Eaton opened the proceedings with a reference to the president's "historic address" on March 12, and argued that the policy that he announced had been "forced upon him and upon the Congress and the Nation by the inescapable facts of history. It is a product of those mysterious and mighty spiritual energies of mankind upon which the world has been, is now and will continue to be carried irresistibly forward to its unpredictable destiny. We can no more escape the impact of these mighty forces flowing beneath the surface of human existence than we can escape the law of gravitation."

The Chairman acknowledged some of the risks inherent in adopting the Truman Doctrine but argued that to not do so would be inviting even "greater risks." He declared himself "profoundly convinced that the safety and security of our Nation and of the world, as well as the very existence of the United Nations," depended on congressional approval of the bill.

The rest of Eaton's statement was a catalog of Soviet aggression and a warning that communism, "in its essence close kin to fascism and nazism, now bestrides the world like a colossus." He warned his colleagues, "No amount of wishful thinking or mawkish sentimentality can disguise the grim facts of the communistic march toward world domination."

Invoking the UN, but in a more martial spirit than his Senate counterpart, he went on: "Much of this expansion of Russian power has taken place in complete disregard of the obligations which she has assumed in the organization of the United Nations. If Russia were permitted to take Greece and Turkey, her next steps of course would be Iran and Afghanistan, India and China. . . ."

The Baptist preacher then reached a rhetorical crescendo: "The supreme issue which confronts us today . . . is simply this: Is the world civilization now in process of creation to be a civilization of freedom or of slavery? Does Americanism or communism hold the key to the future of mankind?"

Perhaps more effective than the Foreign Affairs chairman's oratorical flouirsh was a letter he received from Secretary Marshall, and from which he read. In more restrained prose, Marshall expressed his regret at having to be absent from Washington during the committee's consideration of the bill, and continued, "My strong conviction that the immediate passage of this bill is a matter of the greatest urgency was made even more positive by the recent meeting in Moscow." With his Soviet meetings behind him, there was no further need for diplomatic restraint from the general.

The former Speaker, Sam Rayburn, who had hosted the "Board of Education" meeting during which Truman had received his fateful phone call from the White House, invoked the Almighty in his fervent speech that evening, crying "God help us; God help the world," "if we do not accept our responsibility to aid countries . . . who do not want to be smothered by Communism."

This passionate outburst from the political giant was driven not only by a desire to support the bill, but also because of an ominous danger that had arisen: opponents had seized upon the lack of United Nations involvement to craft an amendment that would gut the bill. Offered by Republican congressman Lawrence H. Smith of Wisconsin, the GOP amendment would delay unilateral action by the United States for sixty days, allowing the UN to take up the issue first. The amendment was designed to appeal not only to outright opponents of the Truman Doctrine, but also to those who supported it in principle but lamented bypassing the UN.

Rayburn was apoplectic, warning his colleagues, "If Greece and Turkey need help, they need it now, not sixty days from now, or a year from now. That might be too late." To support the amendment, he thundered, would be to repeat the mistakes of 1919 and see the Allied triumph of the Second World War "thrown away."

The day of debate ended with the bill's future clouded by the UN amendment's impact on the outcome. Eaton and his lieutenants spent an anxious evening sounding out their colleagues and repeating reassurances regarding the United Nations. The urgency of the crisis in Greece weighed heavily upon the Republican chairman, and like Harry Truman, Eaton believed that time was not a luxury the United States Congress had.

The morning of May 8 was bright and chilly as legislators filed back into the chamber for another contentious day of the debate. As President Truman celebrated his sixty-third birthday at the White House, the members continued to clash over the president's proposed doctrine.

Congressman Smith added what he called a "sugaring" proviso to the UN amendment he had offered the day before: his modified amendment would appropriate $50 to $100 million to the United Nations to allow it to take the lead on Greek aid. In addition, he prepared another amendment to cut the bill's $400 million authorization in half. Action on the amendments was deferred until the following day, while the administration strategized with Eaton. There was little doubt that Smith was doing an effective job trying to push House members into supporting his UN plan.

One of the more colorful of the bill's opponents was Democratic congresswoman Helen Gahagan Douglas of California, a former actress who would be defeated in a Senate race three years later by Congressman Richard M. Nixon. During that campaign, Nixon would shamelessly attack her left-wing political views and dub her "the Pink Lady," saying she was "pink right down to her underwear." Regardless of her undergarments' color, Douglas was determined to stop the Truman Doctrine in its tracks. She offered an amendment that would block any funds for Turkey until the UN was given one year to study the Turkish situation and publish a report.

Undeterred by the skeptical reaction of her colleagues, the California congresswoman offered another amendment, one that would prevent atomic materials and information from being shared with Greece and Turkey. She declared, "We have not spent five minutes inquiring into the beliefs of King Paul of Greece, to whom under this bill the President would have authority to hand over anything or everything related to atomic energy."

One of her bemused colleagues, doubtless having difficulty imagining President Truman handing over atomic materials to the king of Greece, complimented her "well-drawn" amendment, while pointing out its oddity. Republican congressman Jacob K. Javits of New York took a more subtle approach, introducing an amendment that would allow funds for Turkey to be spent but require that the United Nations conduct a review of Soviet pressure on Turkey at the same time. The New York Times reported, "It had been expected widely at the Capitol that Mr. Javits' proposal would receive strong support. It was a phase of the arguments that the United States by its proposed unilateral action, was 'by-passing' the United Nations."

Douglas's amendments could be easily dispatched, but those offered by Javits represented a more significant threat. Fortunately for supporters of the bill, Chairman Eaton was prepared with a response. He had another important letter at the ready, this one from Warren R. Austin, US delegate to the United Nations. Eaton triumphantly read Austin's letter to his colleagues on the House floor:

> In my opinion the United States program for aid to Greece and Turkey does not bypass the United Nations. On the contrary it would be a most essential act in support of the United Nations Charter and would enhance the building of collective security under the United Nations. . . .
>
> The proposed American program will assist in restoring security and stability in Greece and maintaining them in Turkey. When stable conditions are restored in Greece it should be possible to provide such further financial and economic assistance as might then be required through the Economic and Social Council of the United Nations and related specialized agencies.

A letter of support from America's representative to the United Nations was a well-timed blow to the bill's opponents; if the representative of the United States to the UN did not consider the bill a threat to the integrity of that body, then why should Congress? The members responded by clamoring for a vote on the Douglas and Javits amendments; all three were shouted down in a chorus of noes.

The exhausted legislators adjourned late that evening, with a call to reassemble at ten the following morning, May 9. The next day, Eaton again opened the proceedings and contended with more amendments similar to those the House had dispensed with the day before. But he must have taken comfort from the report in the *New York Times* that morning that the bill had "moved into an apparently safe position."

The amendments by Congressman Smith that had caused Eaton and others so much unease were dispatched by large margins. And the endlessly prolonged debates on each amendment that marked the opening phases of the debate yielded to a swifter process, as the bill's opponents realized that they were headed for defeat.

The most elaborate maneuver that day came from Congressman Chet Holifield, a Democrat from California, who at the last moment offered an amendment to recommit the bill and direct the president to work with the United Nations. But like so many others, it was defeated by voice vote. By now, the intention of the House was clear.

The time for debate had ended and a roll call vote had finally begun. The House had spent more time debating the bill than it had any other issue in the crowded first half of 1947—nine hours of general discussion and deliberation over two dozen amendments. As the clerk slowly and monotonously read the members' names, the fate of the Truman Doctrine momentarily hung in the balance. Would members cave to calls of "America's First" and take the path of least resistance by defaulting to a political position that had been the norm for almost two centuries? The air had been charged with passionate denunciations of communism and calls for America to take the lead against it; but opponents of the bill had presented a compelling case as well, warning of the consequences of unlimited worldwide commitments, and reminding their colleagues about the horrors of war. But in the end, Truman's victory was complete. H.R. 2616 passed by a margin of 287 to 187. The vote had been 161 Democrats and 126 Republicans in favor, with 13 Democrats and 93 Republicans voting against.

All that remained was for a conference committee, made up of five members of each house, to reconcile the differences between S. 938 and H.R. 2616. This was swiftly appointed, and over a few days the two bills were

reviewed. Vandenberg and Eaton managed that conference as efficiently as they had the committee process
and the floor debate, and on May 13, the committee reported a final bill. Not surprisingly, it was closer to the
Senate version, with the Vandenberg amendments and
a few minor House provisions retained. The congressional process reached its conclusion in anticlimactic
fashion on May 15, when both the House and the Senate approved the conference report by voice vote. The
Truman Doctrine had passed its legislative test.

At four o'clock that afternoon, the president held his
regular news conference. Reporters jostled one another
in the Oval Office as Truman sat behind his desk,
fielding questions on an array of subjects, including the
final congressional passage of the aid bill. When a journalist asked who would head the Greek mission, the
president demurred, saying that he was still trying to
persuade his favored candidate to take the job. When
asked if the person in question was reluctant, Truman
replied:

> People are always reluctant to take a hard job.
> There's a difference, you know, between doing a pa-
> triotic duty in peacetime and doing a patriotic duty
> when the country is in a shooting war, when there
> is some sort of incentive that makes people more

cooperative, more anxious to make sacrifices—and I can't blame them. But the present period is just as important for the welfare of the country as actual shooting warfare would be. And I hope the patriots will bear that in mind: it is necessary to do things that you don't want to do . . . for the welfare and benefit of your country.

And then, as if to lighten the mood, the president chuckled and said, "I didn't intend to give you a lecture."

The timing was propitious. Just five days after the final bill was approved, the Greek communist leader Nikos Zachariadis met with Stalin in Moscow. No record of their conversation survives, but soon thereafter Stalin gave his official Soviet endorsement to the communists to continue their insurgency against the Greek government. Conscious of the pledge he had made to Churchill three years before, not because of any sense of moral commitment but simply out of caution, Stalin declined to offer any direct aid. But he encouraged the governments of the other Balkan nations in the Soviet orbit to arm and finance the Greek communists.

The White House had reason to be jubilant in the wake of such a sweeping victory. But the responsibilities the United States had now taken upon itself were vast, and this was hardly the time for public celebration. Truman received a reminder of how difficult a choice he had forced Congress to make in the form of a letter from Francis H. Case, a conservative Republican congressman from South Dakota, who wrote in the wake of the House's passage of the bill that dozens of members had supported it only because they feared "pulling the rug out from under you or Secretary Marshall." More ominously, he warned the president, "No country, ours or any other, is wise enough or rich enough, or just plain big enough to run the rest of the world."

The president was close to his ninety-four-year-old mother, Martha Truman, and throughout his presidency visited her as often as he could at her home in Grandview, Missouri, just south of Kansas City. He received word on May 17 that she was gravely ill, and immediately flew to Missouri to see her, confident in the knowledge that legislative work on the aid bill was done and only his signature was required to make it law. There he passed an anxious few days at her bedside, as she seemed for a time on the verge of death. He was still there when the final copy of the legislation was flown out to Kansas City on the evening of May 21.

_____

Most presidents would have signed such a grand and an historic piece of legislation in a lavish White House ceremony, attended by members of Congress and political supporters, with each of the bill's chief sponsors given a pen used for the signing as a memento. But in an understated touch that somehow suited him, President Truman signed the final bill at eight o'clock in the morning on May 22 in the Muehlebach Hotel in Kansas City, Missouri, the same hotel where he would retreat in earlier years for rest and introspection. There was little time for reflection on this particular occasion, however. The *New York Times* described the president as "deeply moved," and the atmosphere as one of "extreme solemnity" as he signed the bill that would become known to history as the Truman Doctrine. The commander in chief sat before a table in the dining room of his suite, surrounded by reporters and photographers and dazzled by the glare of camera lights. Truman asked the assembled newsmen if "everybody was happy," but no trace of a smile crossed his features.

Then he walked a few steps into the adjoining room, and read a brief statement:

The Act . . . is an important step in the building of the peace. Its passage by overwhelming majorities in both Houses of the Congress is proof that the United States earnestly desires peace and is willing to make a vigorous effort to help create conditions of peace.

The conditions of peace include, among other things, the ability of nations to maintain order and independence, and to support themselves economically. In extending the aid requested by two members of the United Nations . . . the United States is helping to further aims and purposes identical with those of the United Nations. Our aid in this instance is evidence not only that we pledge our support to the United Nations but that we act to support it.

With the passage and signature of this Act, our Ambassadors in Greece and Turkey are being instructed to enter into immediate negotiations for agreements which, in accordance with the terms of the Act, will govern the application of our aid. We intend to make sure that the aid we extend will benefit all the peoples of Greece and Turkey, not any particular group or faction.

In a generous and well-deserved tribute to those Republicans and Democrats on Capitol Hill who had shepherded the bill through so quickly, he continued,

"I wish to express my appreciation to the leaders and members of both parties in the Congress for their splendid support in obtaining the passage of this vital legislation."

A *Times* reporter present observed that Truman had completed his prepared statement but seemed to think that something more was called for. He continued,

I want to say also that the press is to be commended—complimented—on the manner in which the program was explained to the country. I think the press made a great contribution toward informing the people of the United States—toward showing just exactly what the intention of the legislation is.

I want to emphasize, too, that this is a step for peace. It is a step to support the United Nations. I cannot emphasize that too strongly. It is set out clearly in the statement, but I am just calling these matters to your attention so that when you read the statement you will know exactly what it means.

President Truman then signed an executive order granting the secretary of state the power to oversee his program. The reporters pressed Truman about whom he would name as the plan's administrator, but he had no answer for them as yet. Those and other challenges could be dealt with later. For now it was time to reflect on the fact that something remarkable had just occurred. There, in a small hotel room in the nation's heartland, the president of the United States had signed a bill that would change America's foreign policy construct for the next seventy years.

Chapter 14
Personnel Is Policy

Truman returned to the White House at the beginning of June, his mother having weathered her medical crisis. During his stay in Missouri, the dutiful son had maintained a near-constant vigil by her bedside, taking time only to sign a series of executive orders. For now, the legislative work was done but the more complicated task of his administration implementing the Truman Doctrine and turning it into a working policy lay ahead.

The State Department prepared a formal agreement with the government of Greece signed in Athens on June 20 by American ambassador Lincoln MacVeagh and Constantine Tsaldaris, deputy prime minister and minister for foreign affairs. The twelve articles of the bilateral agreement outlined America's commitment to

Greece under the terms of the act signed by the president on May 22. The Greek government was forbidden from using any funds from the aid package to make payments on debts to other nations, and the administration was careful to explain Truman would follow the clear intent of the legislation and withdraw aid if the Greek government or the UN Security Council desired. In reality, of course, the Greek government would never ask for the withdrawal aid in the US and Great Britain would veto any attempts by the UN Security Council to make such a request.

Even before the document was signed, the Greek government sent an official note with a deep expression of gratitude:

The hearts of the Greek people are profoundly touched by this proof of the generosity and good will of the American people and of the benevolent interest of a great and friendly nation in the welfare of Greece. The Greek Government on its own behalf and on behalf of the Greek people, wishes to express its deepest appreciation for this magnanimous response to the request of the Greek Government and takes this opportunity to repeat that it turned to the United States for aid only because the devastating results of the war were such as to render impossible the enormous task of reconstruction with the resources remaining to Greece after years of conflict and enemy occupation.

While the agreement was being drafted, there still remained the question of who would administer the program. When former secretary of war and Kansas governor Harry Woodring suggested that a Greek American friend of his help dispense the funds, Truman offered a coarse reply: "I'll keep your friend Leonida in mind but it is my opinion that it would be bad policy to appoint any Greek on a Commission to administer funds for the Greeks. Nearly every Greek in this country belongs to some political faction in Greece and, as you know, there are not any more active connivers and politicans [sic] than are the Greeks."

George McGhee, a Texas oilman and career diplomat, was appointed instead to be the Washington-based coordinator for the aid package. A Rhodes scholar and naval officer, McGhee had served in a number of high-ranking government posts during the Second World War before joining the staff of Undersecretary of State William Clayton. He later recalled the pressure felt being responsible for the allocation of vast amounts of money for projects spread across a distant country; McGhee nonetheless expressed relief that Congress, "after putting up token resistance to appropriating the funds, took little interest in how they would be spent." The Texas oilman made the most of that latitude:

> I well recall the day, after the Greek guerrillas had blown up several of the bridges we had just built, that I quietly, without getting anyone's approval outside of the department, sent the Treasury Department a check transferring $50,000,000 from the Greek economic program to the Defense Department to apply against defeating the guerrillas. No one ever complained or questioned. I concluded that the best way to survive in the Washington bureaucracy was when you had the authority, to "lie low."

Edwin C. Wilson, who had been ambassador to Turkey since 1945, was given control of the aid portfolio for that country. A Florida native and Harvard graduate, Wilson had served as an officer in the army during the First World War. His job in Turkey was far less challenging since the internal state of affairs in that country was more stable; even the Soviet threat had lessened in recent months. As Wilson made clear to the Senate Foreign Relations Committee during his testimony regarding the aid bill, communism held far less appeal in Turkey than Greece. And while Greece had been overrun by the Germans during the war, Turkey had emerged relatively unscathed, reversing its neutrality and entering on the Allied side in the closing months of the war. It was the Greek crisis that had precipitated the Truman Doctrine, with Turkey a secondary (but still vital) concern. Ambassador Wilson was in the enviable position of both dispensing large sums of money and serving as his country's diplomatic representative. In closed session with the Senate committee months before, Acheson had informed members that having a separate aid administrator for Turkey would have brought back unpleasant memories of "foreign courts and foreign officers."

For the equivalent position in Greece, the administration adopted for a different and more cumbersome approach, appointing a separate official to dispense American aid. The president announced his nomination of Dwight P. Griswold as chief of the American Mission for Aid to Greece (AMAG) at a press conference on June 5. Griswold had recently stepped down after three terms as the Republican governor of Nebraska, having been defeated in a bid for the Senate. A native of the state and a graduate of the University of Nebraska, Griswold was an artillery officer in the First World War, a shared experience that may have endeared him to the president. Like Vandenberg, he was also a newspaper editor and publisher turned politician who served in both houses of the state legislature before his election to the governorship. Viewed by many peers as a boorish self-promoter, Griswold was far from an ideal choice, but having relied so heavily on Republican support for passage of the aid bill, Truman was anxious to be accommodating to GOP sensitivities.

A somewhat breathless profile of Griswold by Dana Adams Schmidt nevertheless ran in the *New York Times*. He was declared "the most powerful and very likely the busiest man in Greece," a description that the indefatigable ambassador, Lincoln MacVeagh, must have found galling. It described his task as "trying to patch up the economy of the whole country," and preventing Greek politicians from "turning the $300,000,000 American grant-in-aid into the biggest gravy train ever seen on the Balkan peninsula."

In a seamless blend of sycophancy and condescension, Schmidt referred to Griswold as a "forthright, blunt, and simple man who said what he meant, dropped in the midst of a people to whom deviousness, subtlety and complexity are second nature." The profile also noted approvingly—and somewhat bizarrely—that "his arms have the muscles of a wrestler. . . . Age fifty-three, with a weight of 196 pounds, mostly lean beef, Mr. Griswold could undoubtedly pin any one of the present crop of Greek politicians to the floor with ease." Pulitzer-winning reporting this was not.

By the summer of 1947, the Greek government had fallen. The Americans there concluded that any successor government should be as broad-based as possible, and not dominated again by the country's most extreme right-wing elements.

Griswold was determined to reorder the Greek government to his liking, but his efforts amounted to little. His heavy-handedness offended local sensibilities, and while their need for aid was acute, there was a limit to the amount of political interference they would tolerate. For all the clout that millions of dollars could provide, there remains no substitute for diplomatic finesse, and it took MacVeagh and Loy Henderson, who had been dispatched to Athens from Washington, to help assemble a more balanced Greek cabinet.

Despite his failure, Griswold bragged to Schmidt, "I think what I've done kind of saved the situation. After all, I hold all the cards. When you've got a lot of money to spend you're in a strong position. One fellow wants it spent on the army, another on housing. If you're the fellow who calls the shots, they're a little scared." Privately he complained, "In my judgment we do not need to be affected by a fear that we will be accused of 'interfering.' That accusation will be made even if we do nothing."

The presence of competing American power centers in Athens predictably led to internal conflict among the American diplomatic delegation. Griswold's "rude" behavior, as MacVeagh viewed it, complicated what was already a difficult diplomatic situation. The ambassador was a cultivated and worldly diplomat, educated at Groton and Harvard, and a student of ancient Greek. But his grace and refinement had its limits when challenged by the blunt and unsophisticated head of AMAG—as the keeper of the purse, Griswold held a strong hand despite his personal failings. He presided over a staff of nearly two hundred, divided evenly between civilian and military personnel, and lived with his wife at the King George Hotel. His deputy was his predecessor as governor of Nebraska, R. L. Cochran, who was a road engineer and a military officer before entering politics. He had a politician's practiced touch with the press, as the *Times* profile made clear, and was the dispenser of the American dollars. But while Cochran's presence in Athens may have smoothed over some of Griswold's rougher edges, it did not make conditions in the country any more peaceful.

For all the difficulties experienced in the early stages of the US mission, American aid had a significant impact on Greek economic prospects and the military's ability to stave off the communist insurgency. The BBC's Greece correspondent would later write about those early days of the doctrine.

> American firms of contractors were out on the road, laying down a magic carpet to replace the gashed and pitted surface on which cars had broken their backs and motorists their hearts for long enough; and it was a revelation to see the American overseers at work. In weather which alternated blinding rain with meridian heat, they were up front with the labor gangs, encouraging, demonstrating; and the force of their example, the novelty and brute capability of the tools they used were having an extraordinary effect on the Greek laborers. . . . The new road was leaping forward a mile a day.

By the fall of 1947 MacVeagh's health was worsening, and the ambassador was forced to return home for medical treatment. General Marshall, offended by Griswold's boorish behavior, was determined that the next ambassador to Greece was given the unambiguous authority that MacVeagh lacked. At the secretary of state's request, Truman sent a written order: "The Ambassador is and should be universally recognized as the American representative in Greece charged with dealing with the Greek Government on matters of high policy."

The ever unhelpful Griswold cabled the president in protest: "Either new instructions show that I no longer have your confidence and that of Secretary of State, or else, as I hope, new instructions were based on misconception of situation here and without realization of their practical effect. . . . Under new instructions it would not be possible for me to remain here as I could not do effective or efficient work." Griswold's complaint persuaded the president to put his new directive on hold for the time being, but Truman would have been better served to follow Marshall's instincts and to have taken Griswold's note as a prelude to resignation.

Complicating matters more, rumors of Griswold's political ambitions began to surface across Washington. The administration soon came under pressure by Republicans to name Griswold ambassador, which under other circumstances might have made sense, had he not been so thoroughly undiplomatic in temperament and intellect.

After his health improved, MacVeagh was reassigned to Portugal as ambassador, much to his dismay. As historian David Brewer observed, "It was MacVeagh's tragedy that the clear directive on the ambassador's powers, not drafted until October 1947 and then suspended, was not put firmly into effect that the beginning of his dealings with . . . Griswold."

Griswold himself would eventually resign and return to the United States in August 1948, having used his position in Greece as a platform to reignite his political career. The former govenor achieved his goal of winning a Senate seat in the 1952 elections, but had little time to savor his triumph, dying of a heart attack fifteen months after taking office.

Chapter 15

"You Must Kill All Americans"

In his seminal work *A New Kind of War*, diplomatic historian Howard Jones explains the central dilemma faced by the United States in effectively executing its aid program with Greece: The Greek leadership "understood that, even though they were recipients of American aid, the Truman administration regarded Greece as vital to American security and would give in to many of their demands. In this instance, the tail wagged the dog, making a clash all but inevitable not only between the Greeks and the Americans, but among the Americans themselves."

During the months that Washington politicians were debating the fine points of the Truman Doctrine, Greece was dissolving further into civil war. While the legislation's passage may have heralded a new day in US foreign policy, it did little in the short term to ease the miseries of the Greek people. It soon became apparent that the military dimension of the Truman Doctrine would loom far larger moving foward than the administration had ever suggested to Congress. The deteriorating situation, exacerbated by both relentless guerrilla attacks and a listless Greek response, raised another set of uncomfortable questions in Washington. The stated fears by many isolationist Republicans and progressive Democrats in Congress that American servicemen could end up involved in a foreign civil war quickly proved to be justified. Without direct American involvement, the communist insurgents were looking as if they could still topple the government, with the millions in American aid going to waste.

Truman found himself yet again in a difficult position brought on by the seemingly endless crisis in Greece. He could not deploy American troops to that country in significant numbers without incurring the wrath of Congress and political blowback that could threaten the entire program. But if Washington did nothing to aid the Greek army, an even graver danger loomed. The dilemma was made more complicated by the fact that the United States knew very little about Stalin's intentions; the "Iron Curtain" that Churchill foresaw in his Westminster College speech the previous year was proving to be a nearly impenetrable fortress regarding intelligence gathering. For many in Greece, America's outreach seemed motivated by their own Cold War interests, and not what was best for their own country. One Greek officer observed to an American journalist, "This war in Greece is a battle between the United States and Russia. It happens that it's being fought here. That is our bad luck. But you can't expect us to fight your battle single-handed—at least not with the old spirit."

After months of deliberation, the administration opted to send an advisory and planning group of nearly two hundred military personnel under the authority of the US aid administrator. These officers and enlisted men would advise Greek army leadership on strategy and tactics; they were among the first American "military advisers" that would play a decisive role in future Cold War conflicts. The danger then as well as later was this: What if American personnel were injured or killed during engagements with the enemy? What if their "advisory" capacity turned into something more direct and dangerous?

The risk to these US troops was grave. A captured guerrilla fighter told a *New York Times* reporter in February that he and his compatriots were directed: "You must kill Americans and drive them away because they are the conquerors of Greece. You must shoot Americans whenever they interfere with operations."

The already complicated Greek crisis took a more ominous turn when the communist insurgents, who had already demonstrated extraordinary brutality against the people they were trying to "liberate", adopted a new and disturbing strategy: The kidnapping and removal of thousands of children to camps across the border in Yugoslavia. The Greeks called it *pedhomazoma,* "the stealing of the children." Communist propaganda called it an act of liberation; as Howard Jones relates, "One communist publication contained a story of two exhausted boys over the caption: 'The first picture of persecuted children of democratic Greece shows two of these courageous young comrades who had to flee their homeland before monarchic terror. Entirely without means, they are dependent on international solidarity of people's democratic and progressive countries.'"

The images of young children being torn from their parents and put in camps to be trained as guerrilla fighters inflamed worldwide opinion. The *New York Times* reported in a front-page story that the kidnapping was an attempt to "de-Hellenize Greek children," and "has the long range idea of building up a future

'Slavophone' for Greece indoctrinated with new ide-
ologies." The American government now found itself
in yet another difficult position, under pressure to act
against these savage tactics but unwilling to antago-
nize the Yugoslav government, which might then be
pushed into Stalin's camp. This was an opportunity for
the United Nations to carry out the sort of work for
which it was founded, so the international organization
dispatched inspection teams to the region. The results
were mixed, with information predictably hard to come
by. Many Greeks were fearful of further antagonizing
the guerrillas and would not speak to inspectors, and
others had willingly sent their children away, hoping
to save them from years of chaos brought on by a Nazi
occupation and endless civil war. The vapid, bloodless
UN report ultimately concluded:

(1) A census of children has been taken by the guerrillas in certain areas of Greece under guerrilla
control. The evidence is that this census is in
connexion with the removal of children.

(2) A large number of children has been removed
from certain areas of northern Greece under
guerrilla control to Albania, Bulgaria and Yugoslavia and, according to radio reports from Bel-

grade and Sofia, to certain other countries to the north. However, the Special Committee has not been able to verify, by means available to it, the precise number of children involved.

(3) While a number of parents have agreed under duress to the removal of their children, and some children have in fact been forcibly removed, other parents have consented, or at least failed to object, to such removal. It has not been possible for the Special Committee to determine the exact number of children removed under these categories.

(4) The number of cases reported point to the existence of a programme to remove children from areas of Greece under guerrilla control to certain countries to the north.

(5) Although the responsibility for the initiation of the plan is not known to the Special Committee, it follows from the appearance of Greek children on a large scale in the countries to the north and the numerous announcements of the radios controlled by these Governments that the programme is being carried out with the approval and assistance of these Governments.

The American government feared the truth was far worse, even if the removals did not rise to the level of "genocide" as the Greek government claimed. Truman knew that the only way to save the "liberated" children of Greece, and to bring a degree of stability to the region, was to increase America's focus on the goal of bringing the civil war to an end. While direct military intervention was off the table, a more serious effort to provide leadership and training for the Greek army was essential. A strong leader was needed to change the course of events in Athens, and soon that leader was found.

The appointment of Lieutenant General James A. Van Fleet to lead the Joint United States Military Advisory and Planning Group in Greece proved to be a turning point. Born in New Jersey and raised in Florida, Van Fleet won an appointment to West Point, graduating with the famous class of 1915, the "class the stars fell on," which included Dwight D. Eisenhower, Omar Bradley, and fifty-six other future generals.

As a first lieutenant, Van Fleet joined the 16th Machine Gun Battalion of the 6th Division and saw action on the Western Front in the First World War. Between the wars, he was an instructor and football coach, and like so many of his fellow officers, he found his military career stagnating while serving in the ever-shrinking American army. But the coming of the Second World War saw Van Fleet promoted to colonel and given command of the 8th Infantry Regiment.

The colonel led the 8th Regiment onto Utah Beach on D-Day, and he was soon promoted to brigadier general and given command of the 90th Division, leading his troops across Western Europe and through the German frontier. He ended his wartime service with III Corps, which plunged into the heartland of Nazi Germany and helped capture the Ruhr Valley. He was an imposing six feet two inches tall and 220 pounds, and his quiet competence earned him the respect of his soldiers and superiors. President Truman hailed him as the "finest combat soldier in the U.S. Army," and he was swiftly confirmed for his new position by the Senate on February 18, arriving in Greece less than a week later.

General Van Fleet led a team of 450 military advisers, all of whom were ordered to serve only in an advisory capacity. But the risks of injury and death remained ever present, and the threat of an international incident erupting remained great. Still, the new reality on the ground in Greece was that this large group of US "advisers" were playing a more direct role in bringing the bloody civil war to an end.

Van Fleet initially took a dim view of Greece's military leadership: "Greek commanding officers are intensely politically conscious to the extent that they are reluctant to relieve obviously incompetent officers." Lamenting their "lack of offensive spirit," he sacked many of the more inadequate officers, and worked tirelessly to fashion the force into a more capable fighting force.

The general was a hands-on commander willing to get into the mud with the men he led. His performance proved the perfect rebuttal to the sardonic observations of a British reporter the previous year, who lampooned the inflated expectations the Greeks had of the Americans: "Morale in the army? They would stiffen it with American officers, not sitting on their rumps at a headquarters desk, but slogging it right up there with the front-line fighters." In fact, under Van Fleet, they did just that.

General Van Fleet also pushed for an increase in the number of advisers under his command and streamlined communications with the British. The general's progress with the Greek army was only hampered by political constraints in Washington: The repeated assurances to Congress from the administration that no combat troops would be sent to Greece limited his ability to assign troops to smaller Greek army units that might encounter enemy fighters. Every unarmed American adviser could potentially face grave danger—and was a flash point in what the government still feared could evolve into a proxy war with the Soviet Union.

The American general was well regarded by the Greek government and particularly admired by the king and queen. But his relationship with Ambassador Henry F. Grady—Griswold's successor—was as strained as that of Griswold's with MacVeagh. Grady enjoyed summoning Van Fleet to his office and forcing him to stand and wait as the ambassador sifted through paper on his desk. He chastised the general for spending too much time with Greek royalty, and criticized him in his communications with Washington. Even Truman, who had effusively praised him on his appointment, was now questioning if a replacement was needed "who gets along better with the other American officials and who does not just run a one-man show." But Secretary of War Kenneth C. Royall gave Truman direct (and prescient) advice: "Just leave Van Fleet alone, and—ambassador or no ambassador—he'll win this war."

Meanwhile, the Democratic Army of Greece (DAG) began changing tactics. The guerrillas continued carrying out a blend of their usual unconventional attacks coupled with conventional, large-scale military operations. But the DAG, unsuited for such operations, met repeatedly with failure. This strategic miscalculation in fighting tactics led to the guerrilla's rout in towns across the north of Greece.

DAG leadership then decided to abandon completely their successful guerrilla tactics, and a more concerted effort began to capture and hold territory. In December, the Greek communists declared the establishment of a rival government and promptly assaulted the town of Konitsa, a strategic location near the Albanian border where they sought to establish a rival capital to Athens. But these aggressive advances were met with increased American military aid and Van Fleet's steadying leadership; the Greek National Army was quickly becoming a more professional fighting force. Fortified with American weapons, and fearing the catastrophic consequences of a communist victory, the Greek army fought back with renewed vigor that helped in repulsing the guerrillas. Hundreds were killed on both sides, but the battle proved to be a more stinging defeat for the communists.

Still, the civil war continued to drag on. Greek expectations of what American aid could help them achieve had been exceedingly high, as were American expectations for a swift defeat of guerrilla forces by the Greek government. Short-term disappointment weighed heavily on both the Greek and American governments. Truman faced increased pressure from the left as government officials in Athens executed large numbers of guerrilla fighters and employed other inhumane measures against its enemies. The American journalist John Gunther observed the civil war's personal and political impact:

Let nobody write about Greece lightly. Here is one of the most tragic and painful situations in the world. What is going on in Greece today is real war, though the fighting is desultory and the casualties comparatively light—what is worse, civil war, the most ravaging of all kinds of war. Moreover this is not merely a Greek war but an American war; it is the Americans who make it possible to fight it. Athens is almost like an Anglo-American (mostly American) armed citadel, and neither the Greek army nor government could survive ten days without aid—concrete military aid—from the United States. Not one American citizen in a thousand has any conception of the extent of the American commitment in Greece, the immensity of the American contribution, and the stubborn and perhaps insoluble dilemma into which we—the United States—have plunged ourselves.

Another American journalist's plunge into the Greek Civil War had fatal results that caused further complications for Truman. CBS journalist George Polk, a Texas native and a descendant of President James K. Polk, was a graduate of the University of Alaska and a foreign correspondent. He enlisted in the navy in the Second World War and served as a pilot in the Pacific theater, where he was badly wounded at Guadalcanal. Polk recovered from his injuries and later reported from the war crimes trials in Nuremberg. He was then appointed a radio correspondent for the Middle East, based in Athens.

On May 17, 1948, his bound and blindfolded body was found in Salonika Bay, in northeast Greece. Polk, thirty-four years old, had been shot in the back of the head. He and his young wife were to have left Greece the following week, so that he could take up a new position in Washington. The murdered CBS newsman had been a vociferous critic of the Greek government, shining an uncomfortable spotlight on the corrupt and repressive tactics of government leaders that had driven much of Capitol Hill's opposition to the Truman Doctrine. The Greek government blamed the murder on the communist guerrillas, but most journalists and war observers suspected the government had been involved.

Weeks went by without an arrest. In an attempt to shift suspicion away from themselves, Greek authorities even questioned Polk's widow, suggesting that she had plotted with a lover to kill her husband. Finally, the police arrested a suspect, Gregorios Stachtopolis, who confessed his involvement and named two accomplices, both communists. Another witness testified that the murder had been an elaborate intelligence operation carried out by the guerrillas in order to cast suspicion on the government and convince the Americans to withdraw their aid. Stachtopolis himself testified, "Polk's ghost pursues me night and day. I see him crumpling dead to the bottom of the boat."

Stachtopolis was found guilty and sentenced to life imprisonment, and the United States declared itself satisfied with the outcome. But many covering the war remained convinced that the Greek government had murdered Polk and then perpetrated a cover-up. Howard Jones quotes Polk's CBS colleague Alexander Kendrick, who remembered later, "It was sometimes hard to figure out whether this was a trial for the murder of Polk or a trial of Polk and of other foreign correspondents who have tried to report the Greek story objectively. . . . The most shocking thing about some of the testimony was the undercurrent of feeling that the United States had nothing to complain about, because only one American correspondent had been murdered, whereas several others might have been and perhaps should have been." Stachtopolis served a decade in prison and later recanted his confession, claiming that Greek authorities had forced him to claim responsibility for the murder.

The tragic killing of the CBS newsman was the first of many reminders that fighting in the shadows of the Cold War would exact both financial and moral costs on US leaders executing that strategy. Greece would be the first of many repressive regimes that were both anticommunist and oppressive to basic freedoms within their own countries. Throughout the Cold War, the United States would support a number of regimes that failed to live up to America's stated standards of freedom, but that served as bulwarks against the spread of an even more malignant ideology. Foreign policy is often a choice between several bad options. But the staggering human costs of the Soviets' bloody seventy-four-year reign made the difficult path chosen by Harry Truman the only viable option for America and the Western world.

Chapter 16
Eleven Minutes

Harry Truman faced more consequential crises in his first years in office than any president in US history, with the notable exceptions of George Washington and Abraham Lincoln. Truman spent his first few months guiding the United States to victory over Hitler, making the fateful decision to end the Pacific War through the use of two atomic bombs, and working through the details of a postwar world order. Within two years, his administration would have little choice but to confront Stalin and the Soviets, alleviate the worst refugee crisis in history, and launch unprecedented efforts to keep Western Europe free from communism. The Truman Doctrine, the Marshall Plan, and the formation of NATO were just three of the many policies that helped FDR's successor launch a revolution in US foreign policy. A decision the president would make in 1948 would prove to be equally profound on the shape of world history when Truman decided to recognize Israel's existence following Britain's rapid withdrawal in 1948 from Palestine. In contrast with the Truman administration's unified approach on the Truman Doctrine, the internal debate over Israel was sharp and bitter. Truman found himself in the uncomfortable position of overruling one of his chief advisers, whom he held in the highest regard. Had the same level of dissension been present during the Doctrine's development, Greece and Turkey might well have been abandoned to their fate.

In the late nineteenth and early twentieth centuries, alarmed by growing anti-Semitism on the continent, many European Jews began to emigrate to the United States and the United Kingdom. A smaller number, wishing to return to the land ruled by Jews for several centuries until their expulsion by the Romans, went to Palestine. The ancient Jewish homeland was then a possession of the Ottoman Empire, and within it lived hundreds of thousands of Arabs scattered in villages throughout the territory. Many Palestinian residents understandably viewed the Jewish influx with alarm.

Palestine had been a contested battleground in the First World War. The Ottoman Empire, centered in

Turkey, had allied itself with the Germans, when the British invaded and occupied Palestine in 1917. In an effort to solicit Arab support against the Turks, the British held out the promise of a united Arab state in the region.

But Jewish immigrants to the area were growing in number and winning the support of Jews in Britain and around the world for the creation of a Jewish state in Palestine. This movement, known as Zionism, was endorsed by Winston Churchill, who as early as 1908 declared himself in "full sympathy" with the creation of such a state. The leader of the movement was Chaim Weizmann, a Russian-born Jew and noted chemist who emigrated to England and lobbied the government on behalf of a Jewish homeland. He grew close to the foreign secretary (and former Conservative prime minister), Arthur Balfour; their friendship would have a profound effect on the Middle East's future.

The British government briefly considered the establishment of a Jewish homeland in Uganda, to avoid antagonizing the Arabs in Palestine. Weizmann rejected this, asking Balfour: "Would you give up London to live in Saskatchewan?" The foreign secretary replied by saying that the English people had always lived in London. To this, Weizmann responded, "Yes, and we lived in Jerusalem when London was still a marsh."

The Zionist movement continued growing in influence and counted among its supporters some of the most prominent figures in British public life. One of them was Baron Lionel Walter Rothschild, heir to the vast banking fortune and for more than a decade a member of Parliament. A friend of Weizmann's, he lent his name and resources to the campaign.

In a letter to Lord Rothschild on November 2, 1917, Balfour announced: "His Majesty's Government view with favour the establishment in Palestine of a national home for the Jewish people." This momentous document, soon to become known as the Balfour Declaration, would have consequences far beyond those envisioned by its author. The still-mighty British Empire had pledged itself to create a home for the Jews, but few could have foreseen the founding—two decades later—of a small but powerful nation that the US government would view as a bulwark of Western interests in an authoritarian Middle East.

At the conclusion of the Great War, the League of Nations assigned Palestine to the care of the United Kingdom. This "mandate" was to last throughout the interwar period and the even more terrible war to follow. But the same retrenchment that led Britain to withdraw aid from Greece and Turkey after the war meant that this British mandate would soon come to an end. In 1947, the new United Nations would decide the fate of Palestine, and President Truman, already consumed with crises around the world, would play an enormous role in what happened next. Jews continued moving into Palestine and their Zionist dream began looking more like a possibility than ever before.

On November 29, 1947, amid much controversy, the United Nations announced the "partition" of Palestine into two states, one for Jews and the other for Arabs already living in the country. Truman had lobbied quietly for this partition, despite opposition from the Arab states, the British, and his own State Department. He wrote later of his belief that partition "could open the way to peaceful collaboration between the Arabs and the Jews." Six months later, the British formally withdrew, and the partition went into effect in May 1948.

Jews around the world rejoiced, but Arab leaders were understandably enraged and threatened war. Despite his support for partition and sympathy for the plight of Jews, Truman was cautious about offering public support for Zionism. Given the growing tension in the region, he thought it was in America's best interest for their president to be seen as an honest broker in the conflict. Truman even refused to meet Chaim Weizmann and other Jewish leaders in the White House despite Weizmann's urgent pleas, and their previously warm relationship.

The American president was vexed by the intensity of the Jewish campaign, later recalling,

> Individuals and groups asked me, usually in rather quarrelsome and emotional ways, to stop the Arabs, to keep the British from supporting the Arabs, to furnish American soldiers, to do this, that, and the other. I think I can say that I kept the faith in the rightness of my policy in spite of some of the Jews.

As early as 1946, before the crisis became more acute, Truman exploded in a cabinet meeting over his frustration regarding the region. "If Jesus Christ couldn't satisfy the Jews while on earth, how the hell am I supposed to?"

An old friend who played a key role in Truman's earlier life soften his views on the subject. His partner in the haberdashery business, Eddie Jacobson, came to the White House on March 13 to quietly lobby the president on the topic of Israel. He had remained a loyal friend through the long and turbulent years of Truman's ascent, and had never tried to take advantage of his relationship with the president for personal gain. But as a Jew and an admirer of Weizmann, he felt compelled to make the case for a Jewish homeland. Sitting in the Oval Office, he pointed to a small statue of

Andrew Jackson, one of Truman's heroes, and told the president that he admired Weizmann as much as Truman admired Old Hickory. As Jacobson later recalled, he told Truman that Weizmann was "the greatest Jew who ever lived," and urged him not to let the political attacks he had suffered from other Jewish leaders impact his relationship with Weizmann. The appeal to friendship and Jacobson's Jacksonian reference had the desired effect. Truman looked across the desk at his old friend and said, "You win, you baldheaded son-of-a-bitch. I will see him." Truman soon welcomed Weizmann back to the White House—but the Zionist leader was spirited in through the East Wing, and his visit remained off the record. While the White House meeting was "off the books," it proved more than worthwhile for Weizmann, who secured a pledge from the American president that he would continue supporting the partition between Israel and Palestine.

But the State Department's continued steadfast opposition to any further support for the Zionist enterprise made Truman's promise more difficult to keep. Like Britain's Foreign Office, State had an institutional bias in favor of the Arabs, and as a military man, George Marshall was unwilling to risk any action that might threaten oil supplies, especially if a future conflict with the Soviet Union should require them.

Marshall was supported in his position by Loy Henderson and George Kennan, who had played a vital role in State's efforts regarding the Truman Doctrine the previous year. The latter two wrote a memo to the secretary explaining their views regarding Israel, in which they warned:

> As a result of U.S. sponsorship of UN action leading to the recommendation to partition Palestine, U.S. prestige in the Moslem world has suffered a severe blow and U.S. strategic interests in the Mediterranean and Near East have been seriously prejudiced. Our vital interests in those areas will continue to be adversely affected to the extent that we continue to support partition.

State's strategic concerns led to an open conflict between the State Department and the White House. Henderson faced off against the president's aide, Clark Clifford, who was a passionate advocate for the Zionist cause. These two powerful players in Truman's administration, who had worked closely and effectively together just months earlier on Truman's doctrine, were now bitterly opposed.

Truman remained in a state of denial regarding Marshall's position. When Clifford and others warned the president about the secretary of state's grave concerns, he would brusquely respond that Marshall knew how he felt, which may have been true but mattered little in this instance. And when the American ambassador to the United Nations, without warning or presidential approval, called on the General Assembly to reverse its decision regarding partition, the uproar was overwhelming. Jewish groups around the country reacted with fury at the seeming betrayal, and many forecast doom at the polls that November.

Truman himself was furious, writing in his diary:

> This morning I find that the State Dept. has reversed my Palestine policy. The first I know about it is what I see in the papers! Isn't that hell? I'm now in a position of a liar and a double-crosser. I never felt so low in my life. There are people on the 3rd and 4th levels of the State Department who have always wanted to cut my throat. They are succeeding in doing it.

The president suddenly looked weak, dishonest, and not in charge of his own administration. But Chaim Weizmann, recalling Truman's personal pledge at their past meeting, stood firm, writing to Truman, "The choice for our people, Mr. President, is between statehood and extermination. History and providence have placed this issue in your hands, and I am confident that you will yet decide it in the spirit of the moral law."

The internal crisis over Israel's existence finally came to a head in the Oval Office on the afternoon of May 12, when Marshall and his aides confronted Clifford and his colleagues. The president sat behind his desk and allowed his men to have their say. The argument grew heated as Clifford made the case for recognition, telling those assembled, "No matter what the State Department or anybody thinks, we are faced with the actual fact that there is going to be a Jewish state," and that the United States had a moral duty on behalf of the Jews to support it.

Marshall reacted with cold fury, even suggesting in his diary that he would vote against Truman in future elections if the president he currently served continued to support Israel.

The suggestions made by Mr. Clifford were wrong. . . . To adopt these suggestions would have precisely the opposite effect from that intended by Mr. Clifford. The transparent dodge to win a few votes would not in fact achieve this purpose. The great dignity of the office of the President would be seriously diminished. The counsel offered by Mr.

Clifford was based on domestic political considerations, while the problem which confronted us was
international. . . . If the President were to follow
Mr. Clifford's advice and if in the elections I were
to vote, I would vote against the President.

There was nothing else to say. The man whom Truman revered above all others had just delivered a damn-
ing assessment of the president's Israel policy. Though
he rejoiced in having Marshall by his side in all other
political disputes, Truman would now have to choose
between backing the fervent advice of his most trusted
lieutenant and keeping his pledge to Weizmann. The
Department of State or his White House aides had had
their say. Both sides stalked out of the room, with a
fuming Clifford saying later of Marshall that the re-
markably gifted statesman "didn't know his ass from a
hole in the ground."

Israel's fate now rested with the president of the
United States. On his desk in the Oval Office was a
sign that read, "The Buck Stops Here," and as Truman
would later observe, "The President—whoever he is—
has to decide. He can't pass the buck to anybody. No
one else can do the deciding for him. That's his job."
Thus, when the British formally withdrew from Pal-

estine and the Jewish state was declared on May 14, Truman had to make the call. Marshall, at least, had assured him that he would keep his opposition private, but the risk of the secretary of state's grave reservations being leaked to a newspaper reporter of magazine editor was still high. To go back on his word to Weizmann and Jacobson, however, would be anathema to Truman. And he knew that the horror of the Holocaust cried out for redress, and that only a Jewish homeland could provide safe harbor for Jews seeking shelter and safety around the world. How long would Harry Truman agonize over this fateful decision?

It took eleven minutes.

At six p.m., Washington time, the new state was declared in Palestine, and given the name of Israel. It could not survive without the recognition and support of the most powerful country in the world. At 6:11, the White House announced that the United States would do just that by recognizing Israel's existence. The Israeli nation was born, and America was the first to acknowledge it. Chaim Weizmann, who not long before had slipped into the White House unseen, was to be the first president of Israel.

Almost immediately following the declaration, Israel's new neighbors attacked in concert: Iraq, Syria, Egypt, and Jordan invaded in an attempt to deal Zionism an instantaneous death blow. In a remarkable display of military skill and determination against overwhelming odds, made more impressive given the arms embargo maintained by the United States—the nascent Israeli army repulsed the invading countries. The fighting would continue periodically for another forty years until the 1979 Camp David Accords. Because of that diplomatic breakthrough, there would be no major ground war between Israel and its Arab neighbors over the next forty years.

Chapter 17
Cold War Dawn

For all the rising drama surrounding the creation of Israel and their warring Arab neighbors, events in the Balkans remained the Truman administration's top international concern.

Despite Great Britain's withdrawal of economic support, they still maintained a small troop presence in Greece, which provided added insurance against a feared Soviet invasion. Attlee's strapped British government wished to withdraw even this small military presence, which would have been dangerously destabilizing. Another British retreat would force the Greek army to bear an even greater share of the burden and the conflict would be reframed as a showdown solely between America and the communist world. Acheson had assured Congress that American troops would not

be deployed to Greece, but now congressional leaders rightly feared that the threat of British withdrawal
would lead to that US deployment. Postwar demobilization had reduced US troop strength by 90 percent
and any military presence in Greece and Turkey would
strain America's already challenged military readiness.
Even if there were troops to be spared, the risks of such
a deployment were substantial; what if American and
Soviet troops came to blows? The consequences were
too grim to contemplate. That may be why the Joint
Chiefs of Staff initially supported an American military intervention, but demurred after examining more
closely the events on the ground. The chiefs concluded
that in its demobilized state, the US military was not
even in a position to defeat "the combined forces of Albania, Yugoslavia and Bulgaria; most emphatically not
if these countries receive either covert or active Soviet
support."

Then as now, the "special relationship" between
America and Britain was as vital, regardless of Great
Britain's weakened state. Fortunately, swift and effective American diplomacy kept the British committed,
though discussions about increased American military
intervention would persist for a time. Both Truman and
Attlee's governments understood that a British withdrawal and an American entry into the conflict would

have thrown off the delicate balance between economic and military aid that was the heart of the Truman Doctrine, and possibly provoke a Soviet response.

The maintenance of that balance, and the combined British military and American economic presence in the country, began producing the desired effect: Stalin soon began viewing the guerrilla insurgency as hopeless and refused to become more directly involved. By the winter of 1948, the Soviet leader was ready to cut all support for the DAG:

> They have no prospect of success at all. What, do you think that Great Britain and the United States—the United States, the most powerful state in the world—will permit you to break their line of communication to the Mediterranean? Nonsense. And we have no navy. The uprising in Greece must be stopped, and as quickly as possible.

The Truman Doctrine had achieved its central purpose—making Soviet interference in the Greek Civil War politically and economically unwise. Stalin's strategy developed into an exercise in finding vulnerable countries where he coould expand the Soviet's reach without tempting a direct conflict with the United States. America's economic and military aid—not to mention the deployment of American military advisers—took Greece off of the Soviet Union's "wishlist."

But Tito's government, increasingly determined to show its independence from Moscow, maintained support of the Greek insurgents. Thus in the winter of 1949, the agony of Greece continued into yet another year. On February 5, correspondent Anne O'Hare McCormick reported in the *New York Times*, "The Greek war has reached a point where resistance is in danger of collapsing if it is not strengthened." She related the disturbing message the minister of war delivered to parliament: "At no moment since the Communist attacks against Greece started was the situation as critical as it is now." In words that may resonate to US soldiers stationed in Afganistan, a Greek army commander lamented to McCormick, "We are fighting a shadow army whose aim is not to conquer and occupy but to loot, destroy and move on to the next point of attack. We are fighting terror."

The *Times* article noted that without the aid bill, "Greece would long ago have gone the way of countries in the Soviet orbit," but that victory was still elusive. McCormick described the conflict as "a new kind of war . . . against America and American aid." But he concluded that, if "we are prepared for greater costs

and greater risks . . . Greece . . . is a manageable problem. It is not beyond our means and has only to be understood and talked about seriously to be solvable." Of
the Greeks, she wrote admiringly, "No one who has
watched these people through the darkest days of this
bitter winter can believe their fortitude will not carry
them further."

Some of that fortitude was on display later in that
year when in June, the Greek National Army attacked
guerrilla strongholds across northwest Greece. The
operation proved to be a resounding success, in large
part because the army finally adopted the strategy of
surrounding and capturing guerrilla forces, rather than
simply driving them over the border into Yugoslavia
sanctuaries. It was the same strategy that America's
Union army leadership belatedly adopted during the
Civil War, much to the frustration of President Abraham Lincoln, whose grasp of strategy often seemed
superior to that of his generals. A rebel army prevails
by simply surviving; chasing insurgents away to fight
another day proves to be useless.

American aid remained crucial, but the greatest
hope for ending the ongoing warfare was the acrimoni-
ous split between the two communist titans, Stalin and
Tito. The Yugoslav strongman declared in the spring
of 1948, "No matter how much each of us loves the

land of Socialism, the USSR, he can in no case love his country less." Tito had his own regional ambitions that Stalin could not tolerate, and Tito found it wiser to keep a safe distance from the Soviet tyrant, while holding out the possibility of courting US aid. (Dean Acheson, who by this time had become secretary of state, would later write of the American rapprochement with Tito, "It would be bad politics and bad morals to represent him as an ally of the West. He was and would long remain a staunch Communist and the dictator of a police state. This honest attitude, however, raised difficulties in Congress, where Communists belonged to a genus without subordinate species.")

On June 28, 1948, at Stalin's direction, the Cominform (Communist Information Bureau), an international organization of communist parties, expelled Yugoslavia. Displaying more ideological zeal than common sense, the Communist Party of Greece (KKE) took Stalin's side in his split with Tito. Yugoslavia, already tiring of the ongoing struggle and unwilling to fund the guerrillas much longer, retaliated by reducing its support. In July 1949, Tito closed the border with Greece. One guerrilla leader raged, "Tito has joined the imperialist camp. Yugoslavia's policy is treachery." To an Associated Press reporter, Tito "denied Greek communist accusations that he was ever in league with

Greek Government forces. He said the gradual closing of the frontier would 'safeguard the lives of our working men.'" The Greek guerrillas were consequently starved of aid and cut off from their safe haven across the northern border. As one historian of the conflict has observed, "The rank and file of the KKE, and in particular its leaders, were expendable. Without a trace of compunction, Stalin let them go to their doom."

A delicate diplomatic dance ensued. The United States, after initial caution, began making diplomatic overtures to Yugoslavia in a successful attempt to increase Tito's separation from Moscow. At the same time, the US State Department quietly exerted pressure on the Greek government to both refrain from accusing Yugoslavia of complicity in the *pedhomazoma* and ease some of its more repressive policies.

Still the war continued; the Greek army progressed against its enemies by fits and starts. Truman's administration was confident of eventual victory and was already beginning to focus on other Cold War priorities. Pressure was building to reduce the level of aid to Greece and consequently the size of its army since Greece was but one battleground in the growing Cold War. With tensions rising in Berlin and elsewhere, American leadership soon concluded that it was necessary to begin reallocating resources to other foreign hot spots. Despite General Van Fleet's organizational and inspirational leadership through the crisis, it was important to Truman and Congress that he continued to be seen as an observer. Both he and Truman's State Department knew that if the civil war were to end, it would require something not yet witnessed in the war: decisive Greek leadership.

Chapter 18
An American Triumph

General Alexandros Papagos was appointed commander in chief of the struggling Greek army in 1949. The sixty-five-year-old Papagos had been commander in chief during the successful defense of Greece against Italy, but after the German invasion he was interned in a concentration camp. Following the Nazi withdrawal he returned to his home country and resumed his military career. The general's imperious style and demands for absolute authority made the Greek government reluctant to place him in charge, but the ongoing stalemate left them little choice.

Christopher Woodhouse, an Oxford-educated classicist, soldier, and British diplomat who served in Greece during the Second World War and the early phase of the civil war, wrote in his classic *The Struggle for Greece: 1941–1949*: "Papagos was a superlative staff officer, impeccable in logistic planning and exact calculation, a master of the politics and diplomacy of war. . . . His chief asset was his seniority: he could impose his own plans and wishes on both the Greek high command and the allied military missions."

Woodhouse also expressed reservations about Papagos, saying of him that he had "little experience of high command in battle" (a strange observation to make given his success in the Italian war), but still acknowledged that "everyone agreed that his appointment was a decisive contribution to the victory of the National Greek Army."

Papagos wasted no time in putting his stamp on the army, declaring to his men, "There are no obstacles which cannot be overcome for one who wants and is determined to win," and bluntly instructing his officers "to shoot on the spot anyone under his command who showed negligence or faintheartedness." He reorganized the army's leadership and even fended off a brief attempt by the king to persuade him to become the dictator of Greece. It seemed at last that events were finally aligned for victory.

The Truman Doctrine had been proposed by the president and passed by Congress with remarkable speed. But the implementation of the program took much longer; petty squabbles among American officials in Greece wasted valuable time as the country continued to suffer the ravages of civil war. It took until August 1949, after the infusion of millions in aid, the retreat of Stalin, the guidance of General Van Fleet, the leadership of General Papagos, and the closure of the Yugoslav border, for the Truman Doctrine to finally achieve its aim in Greece. The countless meetings in the State, War, and Navy Building, the painstaking composition of the president's address to a joint session of Congress, the vast amount of press coverage, the flurry of committee hearings, the hours of debate, and the bill signing in a Kansas City hotel room all came down to a final clash in the mountains of northern Greece: Operation Torch, a three-phased assault launched in August 1949. The Greek National Army had always enjoyed a numerical advantage over the guerrillas, but for Operation Torch they would bring that advantage to bear in an overwhelmingly effective way.

Torch was preceded by a coordinated attack from the air, pounding guerrilla positions in Grammos and Vitsi near the Albanian border. The Greek air force, using British Spitfires, rained napalm, a weapon of liquid fire, on the fighters below, burning and suffocating them. The United States had dropped napalm on Japan during the latter part of World War II, and would do so extensively during the Vietnam War. Besides killing and incapacitating the enemy on direct contact, the sticky, sulfurous substance sucked oxygen from the air.

There were three phases of the operation. Torch A, launched on August 5, was a diversionary assault on the Grammos region from two directions, and resulted in the capture of high ground by the army and the cutting of the rebel territory between Grammos and Vitsi.

Torch B was an attack on Vitsi from the air on August 10, followed by a ground assault that drove the guerrillas to the Yugoslav border, where instead of passing through on their way back to the fighting, they were imprisoned because of Tito's new policy. Among those who never made it to the border, casualties were high.

Finally, Torch C was an all-out attack on Grammos, where the guerrilla positions had already been weakened by Torch A. This would prove to be the military climax of the war, and among the spectators to this final victory were the king of Greece and General Van Fleet. On August 24, the attack began with a blistering raid by American Helldivers, a fast carrier-based navy dive bomber that came into service in the middle of the Second World War. Manned by Greek pilots, dozens of the bombers swarmed the air space around Grammos. These combined with Spitfires made the aerial assault exceptionally effective and prepared the way for the army's attack. The guerrillas put up a brave resistance under the circumstances, but were simply overwhelmed by Greek troops funded, trained, and kept in the war by US and British support. The Greek National Army had finally vanquished their enemy.

Van Fleet informed his superiors, "Groups are so broken up that an organization on military lines hardly exists; no mining or harassing activities are engaged in and the omnipresent problems [for the guerrillas] are survival from hunger and cold and escape from searching GNA troops." The remaining communist fighters began melting away, escaping into neighboring Albania where they faced an indifferent reception. After a decade of brutal conflict, the civil war ended with a radio announcement on October 16, 1949.

Almost eighty thousand people died in the conflict, and nearly a million lost their homes. Greece's Civil War was every bit as catastrophic for the country as the German occupation had been. Greece had also paid a heavy price for its status as an early battleground in the conflict between America and the USSR. The battered country had become a pawn in an international power struggle involving not only the United States and the Soviet Union, but also of the ever shifting Balkan political landscape. Crushing the communist insurgency had proven to be a more difficult and drawn out enterprise that would have been expected in the heady days following the Americans' arrival. But while the victory took longer than strategists on both sides of the Atlantic expected, American aid staved off certain defeat, drove Stalin off from his prey, and moved Tito to close his border to rebels seeking sanctuary.

The Truman Doctrine in Greece was an unambiguous American triumph. Its mere declaration by the US president and its passage by Congress had been enough to convince Stalin that Greece was no longer worth his effort, although the Americans would not know that for years. Economic aid helped the Greeks rebuild their infrastructure and begin the process of recovery from years of occupation and civil war. The military contribution was even more substantial, allowing for the growth of the Greek National Army and the use of more advanced weaponry. Together, Generals Van Fleet and Papagos formed an effective team that rallied the spirit of the Greek army. American diplomatic finesse had contained the war and prevented it from spreading into neighboring countries. And the administration's resolution not to engage in direct military conflict inside of Greece prevented the United States from being drawn into what might have become another quagmire or regional war.

Napoleon Bonaparte famously declared, "I'd rather have lucky generals than good ones," and there is no doubt Harry Truman was fortunate in the chain of events that unfolded in Greece from 1947 to 1949. The KKE leadership made a disastrous strategic miscalculation when it abandoned guerrilla tactics in favor of conventional warfare and its attempts to hold territory. Even with aid from across the border, the insurgents lacked the manpower and matériel to successfully conduct such operations, while sacraficing their greatest advantage. The split between Tito and Stalin was a further stroke of good fortune, deftly exploited by Washington but caused in large part by internal rivalry inside the communist world.

In the end it was the statesmanship of Harry Truman—a combination of toughness and restraint—that brought victory in Greece and established the United States as a champion of democracy in this new postwar world. For all the imperfections of the Greek government, and other regimes supported by America throughout the Cold War, none remotely compared to the oppressive and brutal nature of the Soviet Union and its Eastern European satellites. The Truman Doctrine proved to be a declaration of resistance against the most murderous regime in world history, and while the failures that followed in the wake of Truman's policy were often tragic and antithetical to American values, anyone who cherishes individual freedom and the expansion of human rights across the world should celebrate its success.

Chapter 19
Aftermath

The Truman Doctrine has shaped US foreign policy and America's role in the world for the last seven decades. It remained the country's strategic vision throughout the Cold War's entirety, and made the defeat of Soviet communism US foreign policy makers' primary goal. Although the Soviet Union was not mentioned by name in the president's March 12, 1947 speech, every member of Congress in that joint session knew the source of the "outside pressures" that threatened free people everywhere.

John Lewis Gaddis, the Cold War's leading scholar, has called the Truman Doctrine and the policies that flowed from it "an attempt, through political, economic, and, later, military means, to achieve a goal largely psychological in nature: the creation of a state of mind among Europeans conducive to the revival of faith in democratic procedures."

The irony of the Truman Doctrine is that it was fashioned as a means of resistance against the spread of Soviet communism, but it was inspired by a communist insurgency in which Stalin quickly lost interest. Combined British and American support for Greece, even before the doctrine, made the Soviet tyrant wary of becoming ensnarled in a civil war erupting in the country. But the Truman administration had every reason to believe Stalin had designs on Greece, just as his omnivorous appetite for expansion targeted the Balkans and Europe. Few historians doubt that a defeated and demoralized Greece would have soon fallen into the Soviet orbit.

The passage of the Greece and Turkey aid packages proved to be a monumental legislative and diplomatic accomplishment. A nation committed since its founding to the principle of nonintervention in foreign affairs had long been disinclined to lead worldwide crusades for freedom. Woodrow Wilson's attempt to do the same following World War I led to personal and political disaster for the Democratic president. But unlike Wilson, Harry Truman understood the political challenges confronting him. Wilson was an Ivy League president and governor before ascending to the White House while Truman was a creature of the Senate. Truman also was wise enough to surround himself with the tough but subtle diplomacy of Dean Acheson and the diplomatic genius of General George Marshall. Republican Arthur Vandenberg also played a vital role in moving the United States beyond its former "Fortress America" approach to foreign affairs. The spread of democracy would continue decades after the Soviet Union's collapse. While the last several years have exposed autocratic impulses both here at home and across the world, the pandemic that ravaged economies and strained health care systems has proven again that countries led by politicians hostile to democratic norms, a free press, and full transparency soon find themselves overwhelmed by historic events.

Many breathtaking developments would follow America's triumph in Greece. The Truman Doctrine soon inspired a series of initiatives on an even grander scale. One of the most far-reaching was the European Assistance Program, a vast network of aid for Europe that saved millions from starvation and set the continent on the road to economic recovery. Despite his support for the Truman Doctrine, and his unrivaled prestige and status, Secretary Marshall did not play a central role in its creation or promotion on

Capitol Hill. But the general's mission to Moscow had brought him face-to-face with the grim, unblinking determination of the Soviet leadership to sow chaos in Europe and across the globe. Marshall would return to Washington angered by Russian intransigence and convinced that only a massive program of aid could shield the shattered European continent from Stalin's malignant designs.

Marshall unveiled the proposal in a speech at Harvard University on June 5, 1947, just two weeks after Truman signed "An Act to Provide for Assistance to Greece and Turkey" into law. In his matter-of-fact style, Marshall declared:

> The truth of the matter is that Europe's requirements for the next three or four years of foreign food and other essential products—principally from America—are so much greater than her present ability to pay that she must have substantial additional help, or face economic, social and political deterioration of a very grave character. . . . The remedy lies in breaking the vicious circle and restoring the confidence of the European people in the economic future of their own countries and of Europe as a whole.

Truman, having already given his name to the doctrine that shaped American foreign policy, wisely ceded the spotlight to Marshall in the campaign to win congressional support of the European Assistance Plan. Despite the president's recent legislative victory, the former general's standing was still unsurpassed in Washington, and thus the "Truman Doctrine" was followed by the "Marshall Plan." The Soviets' foolish decision to not participate in this histroic program proved a tremendous political boost for the bill in Congress, and Truman signed it into law on April 3, 1948. In the most far-reaching act of practical-minded generosity in history, the United States gave $17 billion in aid to the devastated countries of war-torn Europe.

More momentous events would quickly follow.

As tensions between the United States and the Soviet Union continued to rise in 1948, Berlin became a flash point during the first years of the Cold War. The former German capital was well inside Soviet-controlled East Germany but divided between American-controlled West Berlin and the Soviet-dominated east. The Soviets soon revealed their expansionist designs across Eastern Europe and would eventually bring pressure to bear on West Berlin.

Truman and Vandenberg would use this crisis as an opportunity to join forces again. Western democracies would now have to unite in an armed alliance to resist the growing Soviet threat across Europe. The senator worked tirelessly to build support among his colleagues as Acheson—who had returned to the State Department as secretary after Marshall's retirement—made the diplomatic rounds in Western Europe to forge a unified front against Soviet aggressions. Together America and Western Europe fashioned the North Atlantic Treaty, which bound the United States and eleven European nations in a pact of mutual defense against the USSR. The treaty declared, "The Parties agree that an armed attack against one or more of them in Europe or North America shall be considered an attack against them all." The Senate, in yet another extraordinary display of bipartisanship, approved the treaty by a vote of 82–13 in the summer of 1949. Thus was born the North Atlantic Treaty Organization (NATO), an alliance of democracies that kept Soviet leaders in check and eventually brought the Cold War to a peaceful and victorious end.

The ultimate clash between communism and the free world that would unfold during the Truman administration was the Korean War, a conflict far removed from the European arena that had so occupied the architects of the Truman Doctrine. On June 25, 1950, North Korean troops poured over the border into South

Korea, in a calculated act of aggression supported by Moscow. The "Doctrine" may not have originally been conceived as a worldwide umbrella of American protection, extending as far as Asia, but Truman still acted swiftly in response. This time, the United Nations was not only consulted, but the subsequent war was fought under its banner, albeit with an overwhelming preponderance of American soldiers, armaments, and money. For the United States, Korea proved to be an ugly war of attrition, fought without a congressional declaration and without a clear, overriding strategic aim. The war would outlast the Truman administration, ending in an armistice and stalemate in July 1953, ultimately costing nearly thirty-six thousand American lives.

Truman's 1947 speech regarding Soviet aggression in Europe had strategically omitted mention of the USSR, but the doctrine it launched focused on countering the geopolitical ambitions of Moscow. The monolithic view of communism that was later to prevail until President Richard M. Nixon's opening of China had not yet taken hold. Acheson and his team sought to differentiate between the great-power maneuverings of the Soviet Union and the workings of international communism; if a communist-led government demonstrated its independence of the Soviets, it was not necessarily to be considered an adversary. Harvard and Columbia Professor Zbigniew Brzezinski spent much of the 1950s and 1960s writing and teaching on the subect of comparative communism to a generation of students such as future Secretary of State Madeline Albright. The sweeping nature of Truman's speech may have obscured those fine distinctions, but the mandarins in the State Department clung to those subtleties regardless.

A more nuanced view of comparitive communism did not survive the Korean War. The entry of the Chinese into the conflict put an end to any strategy of triangulation until Nixon's opening of China in 1972 and Jimmy Carter's normalization of relations that was launched in 1979 at Brzezinski's farmhouse in McLean, Virginia. While the Soviet Union was not the sponsor of all communist movements across the world, it became American policy to regard communism itself as a fundamental threat to the United States wherever it manifested itself.

From its inception, the Truman Doctrine had prominent critics. George Kennan, who long feared that his writings on containment of Soviet influence were misunderstood by policy makers, considered the doctrine to be "a blank check to give economic and military aid to any area of the world where the Communists show signs of being successful." He later considered Truman's landmark achievement to be little more than an expression of America's failing, a desire to seek "universal . . . doctrines in which to clothe or justify particular actions."

But Kennan would eventually come around to endorse the soundness of Truman's actions. Losing Europe to Soviet domination would be a devastating reversal to the cause of freedom. Echoing the concerns America's founders had regarding human nature, Kennan darkly observed:

> The fact of the matter is that there is a little bit of the totalitarian buried somewhere, way down deep, in each and every one of us. It is only the cheerful light of confidence and security which keeps this evil genius down at the usual helpless and invisible depth. If confidence and security were to disappear, don't think that he would not be waiting to take their place.

The Truman Doctrine did produce darker consequences in the years ahead. What worked in the fields of Europe led to calamitous consequences in the jungles of Vietnam. The Far East would highlight the doctrine's limits; US policy makers' obsession over "falling dominoes" in Asia would soon enough lead to American humiliation. The communists of North Vietnam under Ho Chi Minh proved themselves to be remarkable fighting warriors and a more fierce nationalist force than the guerrillas of Greece. Lyndon Johnson, living uncomfortably in the shadow of his glamorous predecessor, President Kennedy, was advised by the "best and the brightest" whom JFK had assembled. LBJ steadfastly refused to back down in the face of what he saw as communist aggression. His successor, Richard Nixon, was likewise held captive by his inflexible adoption of Truman's doctrine and the high expectations Americans held regarding conflicts with communist rebels. Withdrawing from Vietnam was viewed as politically unsustainable for Kennedy, Johnson, and Nixon. Two decades of tragic foreign policy miscalculations in Southeast Asia led to the deaths of nearly sixty thousand Americans and over two million Vietnamese.

The United States has likewise been drawn into lengthy conflicts in Afghanistan and Iraq, highlighting once again the limits of American power. President George W. Bush would declare: "I believe that God has planted in every heart the desire to live in freedom," only to discover that many citizens in foreign countries yearned more for safety and security. Those who considered the Truman Doctrine to be too idealistic would have been aghast by the sentiments expressed in Bush's Second Inaugural: "It is the policy of the United States to seek and support the growth of democratic movements and institutions in every nation and culture, with the ultimate goal of ending tyranny in our world." This declaration followed the Truman Doctrine to its most extreme conclusion. It was a miscalculation that the forty-third president himself confronted and eventually rectified in the final two years of his presidency.

But the policies of one president cannot be judged by the actions of his successors. The Truman Doctrine, though bold, was a limited exercise in economic and military aid focused on restraining communism in two of the most strategically important countries of Europe following the Second World War. Had those dominoes fallen, other Western European governments would have likely collapsed as well.

The Truman Doctrine was not a declaration of war, but the recognition of a cold war with the Soviet Union that had already begun. It heralded a new era of American involvement in world affairs, but as Truman said in his joint address to Congress, it was also necessary to protect America's colossal wartime investment in World War II. The carnage of that war that raged from 1939 to 1945 was on a scale never before witnessed in human history. Securing the peace that came from Hitler's defeat was worth the price; now the Soviets had to be confronted.

Writing years later, Joseph M. Jones would recall with wonder "American democracy working at its finest, with the executive branch of the government operating far beyond the normal boundaries of timidity and politics, the Congress beyond usual partisanship, and the American people as a whole beyond selfishness and complacency."

Truman, who owed his presidency to the late Franklin Roosevelt, was determined to win the White House in his own right. And the brief burst of cooperation in 1947 that had made the Truman Doctrine a geopolitical reality was soon forgotten in the scramble for the White House.

Despite his foreign policy successes, few gave Truman a chance of being reelected. The president went into his White House run deeply unpopular. He had acquired a reputation for clumsiness and incompetence in domestic affairs, and his preference for surrounding himself in the White House with old friends and political allies from his colorful political past struck many as being beneath the dignity of his office. By 1948, "to err is Truman" became a popular expression, and a *Newsweek* poll of fifty political pundits found that every one of them predicted his defeat.

His 1948 Republican opponent was the popular Governor Thomas E. Dewey of New York, who had been the party's nominee against Roosevelt four years before. Both Vandenberg and Taft had wanted the nomination, but the former was too proud to campaign for it and the latter was too conservative and wooden to gain traction. Despite his lifelong ambitions, "Mr. Republican" would not follow his father into the White House.

Truman's nemesis and predecessor as vice president, Henry Wallace, was the Progressive Party's candidate, threatening to split the party's left flank. Wallace renewed his opposition to the Truman Doctrine and derided his foreign policy programs as a dangerous and headlong rush to destruction.

If that were not challenging enough, Southern Democrats rebelled against the party's limited support for civil rights. Under the "Dixiecrat" banner, they withdrew from the 1948 Democratic Convention and nominated South Carolina governor Strom Thurmond as their candidate. Now the incumbent president faced a party challenge from the right.

The situation seemed hopeless for Truman. Journalists became less concerned with the election than with the future policies of the Dewey administration. But Truman reached deep within himself, as he had during his seemingly hopeless Senate reelection campaign in 1940, and barnstormed across the country. He refused to back away from his belief that American leadership across the world was required, declaring, "We will get peace in the world if the people of the United States back up their bipartisan foreign policy." In another campaign speech he echoed Vandenberg when he said, "We are now the world's leader. . . . That is the reason it is necessary for our political fights to stop at the water's edge."

This ode to bipartisanship did not prevent him from attacking the remaining isolationist elements inside the Republican Party. In speeches near the close of the campaign, he angrily told voters:

The Communists want a Republican administration, because they believe its reactionary policies will lead to confusion and strife upon which Communism thrives. The Communists want us to get out of Europe and Asia. They want us to stop helping European countries. . . . They want us to withdraw and leave the field entirely to them. They know they can never get what they want so long as the Democratic Party remains in control of this Government. But the Communists have real reason to hope that Republican isolationism will exert its pressure within the Republican Party, and, in a period of time, they can take over nation after nation.

Truman's campaigning became even more intense as the election neared. The Democratic president railed against the Republican "do-nothing" Congress, a cynical tactic given the historic degree of cooperation between the administration and Capitol Hill in the realm of foreign affairs. Still, the Republicans never flagged in their efforts to thwart Truman's ambitious domestic plans. On issues related to labor, health care, and the welfare state, the partisan battle waged on endlessly.

Election Day brought one of the most astonishing political upsets in history: Truman won with 303 electoral votes; Dewey won 189 and Thurmond 39. The president prevailed in the popular vote by a margin of 4.5 percent over his Republican opponent. Truman's tireless travels, ferocious rhetoric, and slashing attacks on his opponents bore political fruit. Dewey, by contrast, had pursued a cautious strategy, confident of victory and unwilling to take any chances.

To his eternal delight, the *Chicago Daily Tribune* printed its front page before the polls had closed, with the blazing headline: DEWEY DEFEATS TRUMAN. In one of the most famous photographs in American political history, a beaming Truman holds up the paper in triumph. He had reason to rejoice: Not only had he won a miraculous political victory, but the Democrats had retaken both houses of Congress. The Republican reign on Capitol Hill was over after two short years, and the Democrats reasserted their longstanding dominance in Washington.

On January 20, 1949, the winter sun glinted off the iron dome of the Capitol as Harry Truman raised his hand and took the inaugural oath. Unlike his first presidential oath nearly four years before, there was no mad scramble for the new president, no panic inside the White House, and no sense of impending doom. Instead, Truman felt a level of vindication enjoyed by few politicians before or since. Coatless and hatless in the chill wind, he delivered his inaugural address before a crowd of one hundred thousand gathered before the East Front of the Capitol. The spirit of the Truman Doctrine infused his every word, and unlike his address to Congress twenty-six months before, this one forthrightly named the "false philosophy" that aid to Greece and Turkey was designed to confront: communism. The newly re-elected president contrasted the blessings of democracy, "based on the conviction that man has the moral and intellectual capacity, as well as the inalienable right, to govern himself with reason and justice," with the communist belief that "man is so

weak and inadequate that he is unable to govern himself, and therefore requires the rule of strong masters."

Truman's Inaugural Address made clear that the
American "program for peace and freedom" would
continue with support for the United Nations, the con-
tinuation of the Marshall Plan, and the further implementation of the Truman Doctrine, to "strengthen
freedom-loving nations against the dangers of aggression." And he looked forward to a day when "those
countries which now oppose us will abandon their de-
lusions and join with the free nations of the world in a
just settlement of international differences."

On April 18, 1951, Senator Arthur Vandenberg died.
He had been suffering from lung cancer—the legacy
of a lifetime of cigar smoking—but even as his body
wasted away, the foreign policy giant labored in the
Senate to maintain the bipartisan foreign policy that he
and the president had worked together to promote. As
his strength gave out, the senator returned to Grand
Rapids to spend his final days. On his death, Truman
issued a moving tribute, praising his Republican partner as:

A patriot who always subordinated partisan advantage and personal interest to the welfare of the
Nation.

In his passing the Senate has lost a pillar of strength in whom integrity was implicit in every decision he made and in every vote he cast during a long tenure. . . .

He formed his opinions only after deliberate study of every aspect of every problem that came before him. His courage was fortified by a good conscience. So he had no fear of the consequences to his personal fortunes when the time came for him to differ from men of great power and influence within his own party on the paramount issue of foreign policy.

Of course we know that his independence cost him dearly in everything save honor. But to him his country's welfare, the security of the Nation and a just and enduring peace in a world of free men were above and beyond all other considerations. . . .

But the death of Vandenberg did not end the era of bipartisan foreign policy. His stated belief that "politics stops at the water's edge" was practiced, however imperfectly, throughout the remainder of the Cold War. Together Truman and Vandenberg built the foundations of an American national security policy that would remain in place for seventy years. When the Vietnam War threatened to tear that consensus asunder, as the disaster in Southeast Asia consumed the presidency of Lyndon Johnson, the Democratic Party began questioning the costs of American global leadership. As the 1960s drew to a close, the Cold War became a source of domestic conflict for Democrats and Republicans rather than a unifying crusade.

Forty years were to pass between Truman's inaugural address and the tearing down of the Berlin Wall. The Cold War he dared to wage proved to be a long and violent struggle, fought in the far corners of the earth while in the shadow of nuclear annihilation. But the eventual triumph of the West vindicated Harry Truman's vision declared before Congress in 1947. In the decades between Truman's speech and the collapse of the Iron Curtain, America had, in fact, become Reagan's "city on a hill" and FDR's "arsenal of democracy."

The worldwide prestige enjoyed by Genereal George Marshall—and Truman's unabashed reverence for him—made the secretary of state the ideal spokesman and advocate for the Marshall Plan. But during the dramatic months when the Truman Doctrine was conceived, it was Dean Acheson who proved to be the indispensable man. He could hardly have been more different from his president; the Yale man's patrician manners and forbidding intellect were in sharp contrast to Truman's simple tastes and homespun humor. But together they shaped US foreign policy as few partners in diplomacy ever have.

Americans have spent the past decade questioning their country's role in the world. Their reservations sound much like those made by Republican isolationists and progressive leaders during Truman's time. Decades of underwriting the defense of Western powers and US military misadventures in the Middle East have led Americans and their leaders to again look inward. Donald Trump's foreign policy was framed by his hostility to Western democratic leaders and a bizarre attraction to former KGB agent and current Russian president Vladimir Putin. Trump let pass no opportunity to undermine NATO, a bulwark against Russian aggression since its founding. Trump also, in effect, ceded Syria to Putin, giving Russia its first beachhead in the Middle East since 1973. And his constant attacks on America's most faithful ally during the Cold War, Germany, led to the American president playing into Russia's hands again by withdrawing troops from the country. While Trump's "America First" theme initially struck a nerve with voters, his ignorance of history and lack of diplomatic skill prevented his administration from making progress on any significant foreign policy issue over four years. His open hostility toward democratically elected leaders and his open admiration for autocrats also caused grave damage to America's reputation on the international stage.

The world has undergone profound change since the thirty-third president proclaimed the Truman Doctrine in 1947, and that change became even more vertiginous as we entered the twenty-first century. After four fitful decades, the twilight struggle of the Cold War ended in 1989 with joyful East Germans vaulting over the crumbling Berlin Wall. Soon after, the Soviet Union collapsed under the weight of its own contradictions, pushed to the brink by the powerful rhetoric and increased defense expenditures of Presidents Ronald Reagan and George H. W. Bush. The heady days of the early 1990s were marked by stunning geopolitical milestones like German reunification and confident predictions of a "peace dividend." Some political observers even suggested the Soviet collapse was "the end of history."

But history races forward still, predictably mocking those of us who try to predict its next move. The national security structures so painstakingly assembled by Harry Truman, George Marshall, Dean Acheson, and Arthur Vandenberg may well be due for reconsideration, but their durability and success across the decades should inform our leaders that it is best to proceed forward with great caution. It was, after all, Truman's foreign policy vision that created the world order that saw the United States dominate all rivals economically, diplomatically, militarily, and culturally for over half a century.

Harry Truman was nothing if not decisive. Confronted in 1947 with a historic challenge—whether to extend massive amounts of aid to Greece and Turkey in peacetime, risking conflict with the Soviet Union—he listened to the advice of his gifted subordinates and quickly moved forward on their recommendations. He acted with similar decisiveness in 1945, when confronted with the terrifying choice of whether to end the Pacific War through a bloody invasion of Japan or the use of atomic bombs. On July 26, 1948, Truman issued an executive order desegregating the armed forces:

> It is hereby declared to be the policy of the President that there shall be equality of treatment and opportunity for all persons in the armed services without regard to race, color, religion or national origin. This policy shall be put into effect as rapidly as possible, having due regard to the time required to effectuate any necessary changes without impairing efficiency or morale.

Later that year, once it became evident that the aging and carelessly maintained White House was falling apart around him, he ordered the demolition of most of its interior, and the rebuilding of its historic rooms with concrete and steel. Truman left his successors a gleaming new executive mansion, and a presidency that was enhanced by a new national security construct that would guide the United States to victory in the Cold War.

In his irreverent but insightful biography of Winston Churchill, Boris Johnson (then mayor of London, now British prime minister) wrote of his subject: "He is the resounding human rebuttal to all Marxist historians who think history is the story of vast and impersonal economic forces. . . . One man can make all the difference." As Johnson wrote of Churchill, so it was with Truman. Had the party bosses at the 1944 Democratic National Convention not intervened to halt the boom for Henry Wallace, and had Wallace become president upon the death of Franklin Roosevelt, there would have been no Truman Doctrine but instead an administration offering little resistance to Soviet designs on the war-ravaged European continent. Rather than a beacon of freedom for oppressed people across the globe, the United States would have likely turned inward again, with consequences impossible to contemplate. Harry S. Truman instead would lead America to a position of supremacy in the world unseen since the days of the Roman Empire, and put the West on a triumphant course in its long, turbulent struggle against Soviet tyranny.

The process that yielded the Truman Doctrine was a breathtaking achievement, and an example of thoughtful bipartisanship, government efficiency, diplomatic brilliance, presidential leadership, and informed public debate. Clark Clifford would call it "one of the proudest moments in American history. What happened during that period was that Harry Truman and the United States saved the free world." Churchill's declaration that Truman saved civilization itself is perhaps the greatest tribute to the thirty-third president, a historical giant dismissed in his time as a strange, little man.

Acknowledgments

A very special thanks to Michael Bishop for his insight, research, and guidance throughout this entire process. Michael proved to be indispensable from start to finish. Thanks as well to Eric Nelson for skillfully managing this project and offering much needed direction and patience when needed the most.

Rachel Campbell, who skillfully manages every project that Mika and I send her way, assembled a great team and guided us toward the finish line. Paige Adams toiled through endless drafts and managed to keep us all organized. Thanks as well to my longtime literary agent Mel Berger at WME, to Ari Emanuel, to Hannah Long at HarperCollins, to the always patient boss and

Mets fan, Phil Griffin, and thanks as well to my *Morning Joe* and *Washington Post* families.

Most importantly, my deepest gratitude goes to Jack, Kate, Andrew, and Joey for enduring the endless book writing, weekly columns, and daily shows. All of my love and admiration belongs to you.

Appendix:

Position and Recommendations of the Department of State Regarding Immediate and Substantial Aid to Greece and Turkey

TOP SECRET

[Washington, undated.]

1.

In view of the wording and timing of the memoranda handed by the British Ambassador to the Secretary of State on February 24 the Department of State regards their presentation as a clear indication that unless the United States is willing to shoulder at once major financial and economic responsibility and a portion of the military responsibility for Greece and to discuss with the British joint measures which should be

taken for the military and economic strengthening of
Turkey:

(a)
Britain will no longer be able to collaborate with us in
joint efforts to hold the line in those countries in order
to prevent a complete collapse which would lay these
countries open to Russian domination.

(b)
In an effort to salvage something of her political positions she may consider herself compelled to pursue
policies of her own with regard to these countries.

2.

In the light of Britain's situation the Department considers this warning to be serious and founded in fact.

3.

The Department considers that this Government has
only this choice: (a) either to accept the general responsibility implied in the British memoranda or (b) to face
the consequences of a widespread collapse of resistance
to Soviet pressure throughout the Near and Middle
East and large parts of western Europe not yet under
Soviet domination or the adverse consequences, from

the standpoint of United States interests, of a possible new British deal with the Russians.

4.

For this reason the Department considers that this Government should accept the responsibilities in question and should do its best to discharge them in such a way as to maintain confidence in the United States and in their own ability to resist Soviet pressure.

5.

The Department considers, however, that before accepting such responsibilities, this Government should obtain satisfactory assurances from the British that we shall have their continued loyal cooperation in our joint efforts to prevent further extension of Soviet power at the expense of the independence of other peoples.

6.

With respect to Turkey, the Department notes that the only specific British proposal thus far is for discussion in the Combined Chiefs of Staff of the strategic and military position of Turkey. The Department recommends that we agree to the immediate undertaking of such discussions and that if as the result of them the Departments of State, War, and Navy find that certain assistance in the form of military supplies is important to the maintenance of Turkish independence, this Government endeavor to furnish an appropriate share.

The British also point out the need for further economic assistance to Turkey. The Department is giving further consideration to this question. It recommends at this time, however, that if in the light of the studies of this Government and after consultation with Great Britain and Turkey the American Government should come to the conclusion that economic and particularly financial assistance to Turkey from abroad is important to the maintenance of Turkish independence, the United States Government endeavor so far as possible to furnish an appropriate share of such assistance under stipulated conditions assuring its most effective utilization.

7.

With respect to Greece, the Department views the problem as falling into two parts, military and economic, which will require separate consideration and treatment. The Department's tentative views on these points are as follows:

(a)

Military

The Department recommends that the United States Joint Chiefs immediately enter into conversations in the Combined Chiefs of Staff, as suggested in paragraph 9 of the British memorandum, with regard to the various military questions involved, and that if as a result of these conversations the Departments of State, War, and Navy should come to the conclusion that Greece must have continued assistance from abroad in the form of military supplies if it is to maintain its independence and restore domestic tranquility, the United States Government so far as it is able furnish an appropriate share of such supplies.

(b)

Economic

It is the view of the Department that the charges upon this Government involved in the assumption of military responsibility in Greece may continue indefinitely unless economic reconstruction in Greece is assured. The need for external assistance is unquestioned, though its exact magnitude cannot now be specified except to assert that such need is considerable. The British estimates as to the total external assistance which will be required to support both military and civilian programs need to be checked.

If this Government is to provide immediate financial assistance to Greece, U.S. interests can be adequately served only by establishing immediately the controls necessary to assure the effective utilization of such assistance. The Greek Government cannot itself provide these controls in the near future.

Nor is it possible for the Greek Government, as now organized and administered, to undertake by itself the detailed and systematic program of restoration required to make the Greek economy self-supporting within the near future. This consideration seems to the Department to call for the establishment of an American Administrative Organization to undertake Greek rehabilitation. Such an Organization should have wide powers over Greek economic life. The establishment and operation of such an Organization would call for a considerable additional outlay in American funds and in American personnel of the highest competence and personal integrity. The Department would expect that such an Organization would be terminated as soon as its services were no longer required.

8.

The Department considers that the program set forth in paragraphs 6 and 7, if put into effect promptly and in its entirety, offers a reasonable chance of success. Half-way measures will not suffice and should not be attempted. They would result merely in the waste of American money and manpower.

9.

The Department recommends that the above program, if agreed to by the Secretaries of War and Navy, be submitted immediately by the three Secretaries to the President for his approval.

10.

If the President's approval is forthcoming, the Department considers that the following further steps should then be taken:

(a)
An appropriate reply be made to the British Government and the specific assurances mentioned above be sought, and this Government at once propose top secret conversations at a high level in regard to the whole

international situation with a view to ascertaining British capabilities and intentions.

(b)
The Secretary of the Treasury, the Secretary of Commerce, and various other members of the Cabinet be informed of this decision and the Secretary of the Treasury be invited to arrange for Treasury participation in any future discussions bearing on financial assistance. Steps be taken by the Administration to obtain the wholehearted support of all other interested executive agencies of this Government in executing the program outlined.

(c)
Every effort be made at the highest governmental level to find means, without waiting for legislation, to alleviate the present Greek financial situation.

(d)
That steps be taken to see that the Greek Government requests at once in a formal manner the assistance of this Government in the rehabilitation of its economic life.

(e)

The problem be discussed privately and frankly by the leaders of the administration with appropriate members of Congress.

(f)

Legislation be drafted, in the light of these discussions with members of Congress and of the findings of the Combined Chiefs of Staff, and this legislation be submitted to Congress; such legislation might well include authorization for the President under certain conditions within prescribed limits to extend loans, credits, or grants to Greece and/or Turkey; also for the transfer to Greece or Turkey or both of military supplies not transferable under existing law; and any necessary authorization for the supply of personnel.

(g)

In the meantime measures be taken immediately to transfer to Greece such available military equipment and other supplies as the three Departments find are urgently needed by Greece and are transferable under existing legislation.

(h)

Measures be adopted to acquaint the American public with the situation and with the need for action along the proposed lines.

PRESIDENT HARRY S. TRUMAN'S ADDRESS BEFORE A JOINT SESSION OF CONGRESS, MARCH 12, 1947

Mr. President, Mr. Speaker, Members of the Congress of the United States:

The gravity of the situation which confronts the world today necessitates my appearance before a joint session of the Congress. The foreign policy and the national security of this country are involved.

One aspect of the present situation, which I wish to present to you at this time for your consideration and decision, concerns Greece and Turkey.

The United States has received from the Greek Government an urgent appeal for financial and economic assistance. Preliminary reports from the American Economic Mission now in Greece and reports from the American Ambassador in Greece corroborate the statement of the Greek Government that assistance is imperative if Greece is to survive as a free nation.

I do not believe that the American people and the

Congress wish to turn a deaf ear to the appeal of the Greek Government.

Greece is not a rich country. Lack of sufficient natural resources has always forced the Greek people to work hard to make both ends meet. Since 1940, this industrious and peace loving country has suffered invasion, four years of cruel enemy occupation, and bitter internal strife.

When forces of liberation entered Greece they found that the retreating Germans had destroyed virtually all the railways, roads, port facilities, communications, and merchant marine. More than a thousand villages had been burned. Eighty-five percent of the children were tubercular. Livestock, poultry, and draft animals had almost disappeared. Inflation had wiped out practically all savings.

As a result of these tragic conditions, a militant minority, exploiting human want and misery, was able to create political chaos which, until now, has made economic recovery impossible.

Greece is today without funds to finance the importation of those goods which are essential to bare subsistence. Under these circumstances the people of Greece cannot make progress in solving their problems of reconstruction. Greece is in desperate need of financial and economic assistance to enable it to resume purchases of food, clothing, fuel and seeds. These are indispensable for the subsistence of its people and are obtainable only from abroad. Greece must have help to import the goods necessary to restore internal order and security, so essential for economic and political recovery.

The Greek Government has also asked for the assistance of experienced American administrators, economists and technicians to insure that the financial and other aid given to Greece shall be used effectively in creating a stable and self-sustaining economy and in improving its public administration.

The very existence of the Greek state is today threatened by the terrorist activities of several thousand armed men, led by Communists, who defy the government's authority at a number of points, particularly along the northern boundaries. A Commission appointed by the United Nations Security Council is at present investigating disturbed conditions in northern Greece and alleged border violations along the frontier between Greece on the one hand and Albania, Bulgaria, and Yugoslavia on the other.

Meanwhile, the Greek Government is unable to cope with the situation. The Greek army is small and poorly equipped. It needs supplies and equipment if it is to restore the authority of the government throughout Greek territory. Greece must have assistance if it is to become a self-supporting and self-respecting democracy.

The United States must supply that assistance. We have already extended to Greece certain types of relief and economic aid but these are inadequate.

There is no other country to which democratic Greece can turn.

No other nation is willing and able to provide the necessary support for a democratic Greek government.

The British Government, which has been helping Greece, can give no further financial or economic aid after March 31. Great Britain finds itself under the necessity of reducing or liquidating its commitments in several parts of the world, including Greece.

We have considered how the United Nations might assist in this crisis. But the situation is an urgent one requiring immediate action and the United Nations and its related organizations are not in a position to extend help of the kind that is required.

It is important to note that the Greek Government has asked for our aid in utilizing effectively the financial and other assistance we may give to Greece, and in improving its public administration. It is of the utmost importance that we supervise the use of any funds made available to Greece; in such a manner that each dollar spent will count toward making Greece self-supporting, and will help to build an economy in which a healthy democracy can flourish.

No government is perfect. One of the chief virtues of a democracy, however, is that its defects are always visible and under democratic processes can be pointed out and corrected. The Government of Greece is not perfect. Nevertheless it represents 85 percent of the members of the Greek Parliament who were chosen in a recent election. Foreign observers, including 692 Americans, considered this election to be a fair expression of the views of the Greek people.

The Greek Government has been operating in an atmosphere of chaos and extremism. It has made mistakes. The extension of aid by this country does not mean that the United States condones everything that the Greek Government has done or will do. We have condemned in the past, and we condemn now, extremist measures of the right or the left. We have in the past advised tolerance, and we advise tolerance now.

Greece's neighbor, Turkey, also deserves our attention.

The future of Turkey as an independent and economically sound state is clearly no less important to the freedom-loving peoples of the world than the future of

Greece. The circumstances in which Turkey finds itself today are considerably different from those of Greece. Turkey has been spared the disasters that have beset Greece. And during the war, the United States and Great Britain furnished Turkey with material aid.

Nevertheless, Turkey now needs our support.

Since the war Turkey has sought financial assistance from Great Britain and the United States for the purpose of effecting that modernization necessary for the maintenance of its national integrity.

That integrity is essential to the preservation of order in the Middle East.

The British government has informed us that, owing to its own difficulties can no longer extend financial or economic aid to Turkey.

As in the case of Greece, if Turkey is to have the assistance it needs, the United States must supply it. We are the only country able to provide that help.

I am fully aware of the broad implications involved if the United States extends assistance to Greece and Turkey, and I shall discuss these implications with you at this time.

One of the primary objectives of the foreign policy of the United States is the creation of conditions in which we and other nations will be able to work out a way of life free from coercion. This was a fundamental issue in the war with Germany and Japan. Our victory was won over countries which sought to impose their will, and their way of life, upon other nations.

To ensure the peaceful development of nations, free from coercion, the United States has taken a leading part in establishing the United Nations, The United Nations is designed to make possible lasting freedom and independence for all its members. We shall not realize our objectives, however, unless we are willing to help free peoples to maintain their free institutions and their national integrity against aggressive movements that seek to impose upon them totalitarian regimes. This is no more than a frank recognition that totalitarian regimes imposed on free peoples, by direct or indirect aggression, undermine the foundations of international peace and hence the security of the United States.

The peoples of a number of countries of the world have recently had totalitarian regimes forced upon them against their will. The Government of the United States has made frequent protests against coercion and intimidation, in violation of the Yalta agreement, in Poland, Rumania, and Bulgaria. I must also state that in a number of other countries there have been similar developments.

At the present moment in world history nearly every nation must choose between alternative ways of life. The choice is too often not a free one.

One way of life is based upon the will of the majority, and is distinguished by free institutions, representative government, free elections, guarantees of individual liberty, freedom of speech and religion, and freedom from political oppression.

The second way of life is based upon the will of a minority forcibly imposed upon the majority. It relies upon terror and oppression, a controlled press and radio; fixed elections, and the suppression of personal freedoms.

I believe that it must be the policy of the United States to support free peoples who are resisting attempted subjugation by armed minorities or by outside pressures.

I believe that we must assist free peoples to work out their own destinies in their own way.

I believe that our help should be primarily through economic and financial aid which is essential to economic stability and orderly political processes.

The world is not static, and the status quo is not sacred. But we cannot allow changes in the status quo in violation of the Charter of the United Nations by such methods as coercion, or by such subterfuges as political infiltration. In helping free and independent nations to maintain their freedom, the United States will be giving effect to the principles of the Charter of the United Nations.

It is necessary only to glance at a map to realize that the survival and integrity of the Greek nation are of grave importance in a much wider situation. If Greece should fall under the control of an armed minority, the effect upon its neighbor, Turkey, would be immediate and serious. Confusion and disorder might well spread throughout the entire Middle East.

Moreover, the disappearance of Greece as an independent state would have a profound effect upon those countries in Europe whose peoples are struggling against great difficulties to maintain their freedoms and their independence while they repair the damages of war.

It would be an unspeakable tragedy if these countries, which have struggled so long against overwhelming odds, should lose that victory for which they sacrificed so much. Collapse of free institutions and loss of independence would be disastrous not only for them but for the world. Discouragement and possibly failure would quickly be the lot of neighboring peoples striving to maintain their freedom and independence.

Should we fail to aid Greece and Turkey in this fateful hour, the effect will be far reaching to the West as well as to the East.

We must take immediate and resolute action.

I therefore ask the Congress to provide authority for assistance to Greece and Turkey in the amount of $400,000,000 for the period ending June 30, 1948. In requesting these funds, I have taken into consideration the maximum amount of relief assistance which would be furnished to Greece out of the $350,000,000 which I recently requested that the Congress authorize for the prevention of starvation and suffering in countries devastated by the war.

In addition to funds, I ask the Congress to authorize the detail of American civilian and military personnel to Greece and Turkey, at the request of those countries, to assist in the tasks of reconstruction, and for the purpose of supervising the use of such financial and material assistance as may be furnished. I recommend that authority also be provided for the instruction and training of selected Greek and Turkish personnel.

Finally, I ask that the Congress provide authority which will permit the speediest and most effective use, in terms of needed commodities, supplies, and equipment, of such funds as may be authorized.

If further funds, or further authority, should be needed for purposes indicated in this message, I shall not hesitate to bring the situation before the Congress. On this subject the Executive and Legislative branches of the Government must work together.

This is a serious course upon which we embark.

I would not recommend it except that the alternative is much more serious. The United States contributed $341,000,000,000 toward winning World War II. This is an investment in world freedom and world peace.

The assistance that I am recommending for Greece and Turkey amounts to little more than one-tenth of 1 percent of this investment. It is only common sense that we should safeguard this investment and make sure that it was not in vain.

The seeds of totalitarian regimes are nurtured by misery and want. They spread and grow in the evil soil of poverty and strife. They reach their full growth when the hope of a people for a better life has died. We must keep that hope alive.

The free peoples of the world look to us for support in maintaining their freedoms.

If we falter in our leadership, we may endanger the peace of the world—and we shall surely endanger the welfare of our own nation.

Great responsibilities have been placed upon us by the swift movement of events.

I am confident that the Congress will face these responsibilities squarely.

PUBLIC LAW 75, 80TH CONGRESS
AN ACT

To provide for assistance to Greece and Turkey

Whereas the Governments of Greece and Turkey have sought from the Government of the United States immediate financial and other assistance which is necessary for the maintenance of their national integrity and their survival as free nations; and

Whereas the national integrity and survival of these nations are of importance to the security of the United States and of all freedom-loving peoples and depend upon the receipt at this time of assistance; and

Whereas the Security Council of the United Nations has recognized the seriousness of the unsettled conditions prevailing on the border between Greece on the one hand and Albania, Bulgaria, and Yugoslavia on the other, and, if the present emergency is met, may subsequently assume full responsibility for this phase of the problem as a result of the investigation which its commission is currently conducting; and

Whereas the Food and Agriculture Organization mission for Greece recognized the necessity that Greece receive financial and economic assistance and recommended that Greece request such assistance from the appropriate agencies of the United Nations and from the Governments of the United States and the United Kingdom; and

Whereas the United Nations is not now in a position to furnish to Greece and Turkey the financial and economic assistance which is immediately required; and

Whereas the furnishing of such assistance to Greece and Turkey by the United States will contribute to the freedom and independence of all members of the United Nations in conformity with the principles and purposes of the Charter: Now, therefore,

Be it enacted by the Senate and House of Representatives of the United States of America in Congress assembled, That, notwithstanding the provisions of any other law, the President may from time to time when he deems it in the interest of the United States furnish assistance to Greece and Turkey, upon request of their governments, and upon terms and conditions determined by him—

1. (1) by rendering financial aid in the form of loans, credits, grants, or otherwise, to those countries;

2. (2) by detailing to assist those countries any persons in the employ of the Government of the United States; and the provisions of the Act of May 25, 1938 (52 Stat. 442), as amended, applicable to personnel detailed pursuant to such Act, as amended, shall be applicable to personnel detailed pursuant to this paragraph: Provided, however, That no civilian personnel shall be assigned to Greece or Turkey to administer the purposes of this Act until such personnel have been investigated by the Federal Bureau of Investigation;

3. (3) by detailing a limited number of members of the military services of the United States to assist those countries, in an advisory capacity only; and the provisions of the Act of May 19, 1926 (44 Stat. 565), as amended, applicable to personnel detailed pursuant to such Act, as amended, shall be applicable to personnel detailed pursuant to this paragraph;

4. (4) by providing for (A) the transfer to, and the procurement for by manufacture or otherwise and the transfer to, those countries of any articles, services, and information, and (B) the instruction and training of personnel of those countries; and

5. (5) by incurring and defraying necessary expenses, including administrative expenses and expenses for compensation of personnel, in connection with the carrying out of the provisions of this Act.

SEC. 2. (a) Sums from advances by the Reconstruction Finance Corporation under section 4 (a) and from the appropriations made under authority of section 4 (b) may be allocated for any of the purposes of this Act to any department, agency, or independent establishment of the Government. Any amount so allocated shall be available as advancement or reimbursement, and shall be credited, at the option of the department, agency, or independent establishment concerned, to appropriate appropriations, funds or accounts existing or established for the purpose.

(b) Whenever the President requires payment in advance by the Government of Greece or of Turkey for assistance to be furnished to such countries in accordance with this Act, such payments when made shall be credited to such countries in accounts established for the purpose. Sums from such accounts shall be allocated to the departments, agencies, or independent establishments of the Government which furnish the assistance for which payment is received, in the same manner, and shall be available and credited in the same manner, as allocations made under subsection (a) of this section. Any portion of such allocation not used as reimbursement shall remain available until expended.

(c) Whenever any portion of an allocation under subsection (a) or subsection (b) is used as reimbursement, the amount of reimbursement shall be available for entering into contracts and other uses during the fiscal year in which the reimbursement is received and the ensuing fiscal year. Where the head of any department, agency, or independent establishment of the Government determines that replacement of any article transferred pursuant to paragraph (4) (A) of section 1 is not necessary, any funds received in payment therefor shall be covered into the Treasury as miscellaneous receipts.

(d) (1) Payment in advance by the Government of Greece or of Turkey shall be required by the President for any articles or services furnished to such country under paragraph (4) (A) of section 1 if they are not paid for from funds advanced by the Reconstruction Finance Corporation under section 4 (a) or from funds appropriated under authority of section 4 (b).

(2) No department, agency, or independent establishment of the Government shall furnish any articles or services under paragraph (4) (A) of section 1 to ei-

ther Greece or Turkey, unless it receives advancements or reimbursements therefor out of allocations under subsection (a) or (b) of this section.

SEC. 3. As a condition precedent to the receipt of any assistance pursuant to this Act, the government requesting such assistance shall agree (a) to permit free access of United States Government officials for the purpose of observing whether such assistance is utilized effectively and in accordance with the undertakings of the recipient government; (b) to permit representatives of the press and radio of the United States to observe freely and to report fully regarding the utilization of such assistance; (c) not to transfer, without the consent of the President of the United States, title to or possession of any article or information transferred pursuant to this Act nor to permit, without such consent, the use of any such article or the use or disclosure of any such information by or to anyone not an officer, employee, or agent of the recipient government; (d) to make such provisions as may be required by the President of the United States for the security of any article, service, or information received pursuant to this Act; (e) not to use any part of the proceeds of any loan, credit, grant, or other form of aid rendered pursuant to this Act for the making of any payment on account of the principal or interest on any loan made to such government by any other foreign government; and (f) to give full and continuous publicity within such country as to the purpose, source, character, scope, amounts, and progress of United States economic assistance carried on therein pursuant to this Act.

SEC. 4. (a) Notwithstanding the provisions of any other law, the Reconstruction Finance Corporation is authorized and directed, until such time as an appropriation shall be made pursuant to subsection (b) of this section, to make advances, not to exceed in the aggregate $100,000,000, to carry out the provisions of this Act, in such manner and in such amounts as the President shall determine.

(b) There is hereby authorized to be appropriated to the President not to exceed $400,000,000 to carry out the provisions of this Act. From appropriations made under this authority there shall be repaid to the Reconstruction Finance Corporation the advances made by it under subsection (a) of this section.

SEC. 5. The President may from time to time prescribe such rules and regulations as may be necessary and proper to carry out any of the provisions of this Act; and he may exercise any power or authority conferred upon him pursuant to this Act through such department, agency, independent establishment, or officer of the Government as he shall direct.

The President is directed to withdraw any or all aid authorized herein under any of the following circumstances:

(1). If requested by the Government of Greece or Turkey, respectively, representing a majority of the people of either such nation;

(2). If the Security Council finds (with respect to which finding the United States waives the exercise of any veto) or the General Assembly finds that action taken or assistance furnished by the United Nations makes the continuance of such assistance unnecessary or undesirable;

(3). If the President finds that any purposes of the Act have been substantially accomplished by the action of any other inter-governmental organizations or finds that the purposes of the Act are incapable of satisfactory accomplishment: and

(4). If the President finds that any of the assurances given pursuant to section 3 are not being carried out.

SEC. 6. Assistance to any country under this Act may, unless sooner terminated by the President, be terminated by concurrent resolution by the two Houses of the Congress.

SEC. 7. The President shall submit to the Congress quarterly reports of expenditures and activities, which shall include uses of funds by the recipient governments, under authority of this Act.

SEC. 8. The chief of any mission to any country receiving assistance under this Act shall be appointed by the President, by and with the advice and consent of the Senate, and shall perform such functions relating to the administration of this Act as the President shall prescribe.

Approved May 22, 1947

A Note on Sources

These books and archival records are a sampling of materials consulted by the author, and are recommended to those who wish to learn more about this critical period of American postwar history.

BIOGRAPHIES OF TRUMAN

The best-known and most readable biography of the thirty-third president is the Pulitzer Prize–winning *Truman* by David McCullough (1992). It paints an admiring and lavishly detailed portrait of its subject and won the man from Missouri a whole new legion of admirers. A less literary but more scholarly and balanced study is *Man of the People: A Life of Harry S. Truman* by Alonzo Hamby (1995). *The Accidental President:*

Harry S. Truman and the Four Months That Changed the World by A. J. Baime is a gripping account of Truman's first steps onto the presidential stage, although the narrative necessarily ends long before the debut of the doctrine.

THE ADMINISTRATION

The life and presidency of Harry S. Truman have been the subject of countless books and articles, but the Truman Doctrine has received relatively little attention from scholars. So crowded is the canvas of the immediate postwar years, with massive military demobilization, the Berlin Blockade and Airlift, the Marshall Plan, and countless other crises, that the doctrine sometimes gets lost in the background. However, those scholars who have examined the topic have done so thoughtfully and well, and readers wishing to learn more should explore the works mentioned below.

Perhaps the best and most exhaustive scholarly study of the doctrine is *"A New Kind of War": America's Global Strategy and the Truman Doctrine in Greece* by Howard Jones (1989). Jones, a diplomatic historian of note, effectively illuminates the conflict in both Washington and the Balkans over confronting communist aggression.

For a detailed study of the crafting of Truman's speech, see *Proclaiming the Truman Doctrine: The Cold War Call to Arms*, by Denise M. Bostdorff (2008).

A number of the participants in this historical drama left compelling memoirs that were valuable sources in the crafting of this narrative. The most immediate and compelling are those of State Department official Joseph M. Jones, a participant in the proceedings. His *Fifteen Weeks: February 21–June 5, 1947* (1955) vividly conveys the determination of Dean Acheson and his team to act swiftly in response to the crisis and provides important detail and chronology. Acheson's own memoir, *Present at the Creation: My Years in the State Department* (1969) is like the man himself: mordant, witty, and admiring of Truman. The book is a wonderful portrait of a lost age of American diplomacy. (For a more objective look at Acheson, consult *Dean Acheson: A Life in the Cold War* by Robert L. Beisner [2006].) Finally, Truman's own account of his time in office, *Memoirs by Harry S. Truman, Volume 2: Years of Trial and Hope* is a useful, sometimes pungent record of his thoughts and actions during those fateful weeks in 1947.

The Foreign Relations of the United States (FRUS), the official history of American diplomacy published by the Office of the Historian at the Department of State, is an astonishing compendium of documents begun in

1861. Freely available online, it allows the researcher superb insights into the thinking and actions of policy makers.

Various newspaper archives, especially that of the *New York Times,* provide contemporary and often colorful observations. Thankfully, the days of scrolling through illegible microfilm are mostly over, and the online *Times* archive, available by subscription, is compendious and easily searchable.

CONGRESS

For the legislative history of the Truman Doctrine, some of the sources above are helpful, but most valuable of all is the *Congressional Record,* which first appeared in 1873. A detailed record of deliberations on the House and Senate floors, and in committees, it is a priceless resource. Unfortunately, the years covered in this book are not freely available online and must be accessed either through subscription or at research libraries. The author has also relied upon his own experience as a member of the House of Representatives for an insight into the congressional debates on the doctrine.

The sometimes troubled collaboration between Truman and Senator Arthur Vandenberg is thoughtfully explored by Lawrence J. Haas in his *Harry & Arthur: Truman, Vandenberg, and the Partnership That Created the Free World* (2016). For more about Senator Robert A. Taft and his worldview, consult *Mr. Republican: A Biography of Robert A. Taft* (1972).

THE GREEK CIVIL WAR

The classic English-language account of the Greek Civil War is *The Struggle for Greece: 1941–1949* by C. M. Woodhouse (1976). A veteran of Britain's Special Operations Executive (SOE), Woodhouse helped organize Greek resistance forces during the German occupation, and later served as a Conservative member of Parliament and a member of the House of Lords. His experience on the ground gave him a unique insight into the complexity of the bitter Greek conflict. *Greece, the Decade of War: Occupation, Resistance and Civil War* by David Brewer (2016) is a more recent, scholarly study based on new evidence. Another important work is *An International Civil War: Greece, 1943–1949* by the late André Gerolymatos (2016), who was a native of Greece and professor of history at Simon Fraser University in British Columbia.

Crete: The Battle and the Resistance by Antony Beevor (1991) includes a vivid and detailed narrative of the German invasion of Greece. For Churchill's Christmas visit to Athens in 1944, see *Churchill: Walking with Destiny* by Andrew Roberts (2018), and *Winston Churchill: Road to Victory: 1941–1945* (1986), volume 7 of the official biography by Sir Martin Gilbert.

About the Author

JOE SCARBOROUGH is a *New York Times* bestselling author, a *Washington Post* columnist, the creator of *Morning Joe*, and a former United States congressman. He has been named to the TIME 100 list of the world's most influential people.